PROBING SIMPLE VEGETATION DATA FOR COMPLEX STRUCTURAL TRAITS

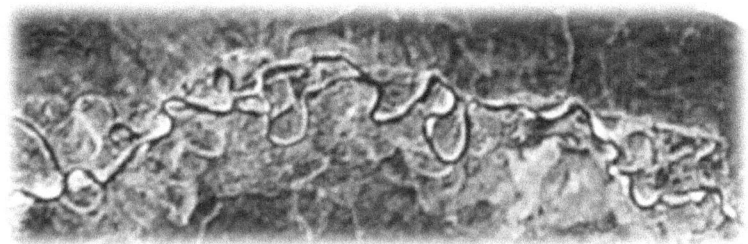

Boron River oxbows in the Aleza Lake area, British Columbia. Google Map 54.156666 N, 122.094047 W.

Front cover: Dimond Ring Lake
A typical peat bog complex (muskeg) in Sub-Boreal Spruce forest. Google Map 54.043601N, 122.008147 W

About the book:
The objective is to outline concepts and introduce the modus operandi of probing multistate phytosociological data by advanced techniques for information potential regarding complex structural traits in the source vegetation.

IN MEMORIAM

Vladimir J. Krajina, Wilfred B. Schofield, Thomas M. C. Taylor
and Ferenc Tuskó

They gave liberally of their time to us in advanced studies at U.B.C.

PROBING SIMPLE VEGETATION DATA FOR COMPLEX STRUCTURAL TRAITS

László Orlóci FRSC, HAS e.m.
Western University, London, Ontario

Technical assistance by
Márta Mihály BSF, DFE

SCADA Publishing – Canada

Refer to this monograph as:
Orlóci, L. 2017. Probing simple vegetation data for complex structural traits. SCADA Publishing, Canada. Online Edition: https://createspace.com/7025180

Look for these monographs:

Orlóci. L. 1916. Statistical quantum ecology. Essays on the resonator complex model of the vegetation stand. SCADA Publishing, Canada. Online Edition: https://createspace.com/ 6509504

Orlóci, L. 2015. Diversity analysis, holistic energetics, and statistics. The resonator complex model of the vegetation stand. SCADA Publishing, Canada. Online edition: https://www.createspace.com/5783923

Orlóci, L. 2015. Energy-based vegetation mapping. A case study in statistical quantum ecology. SCADA Publishing, Canada. Online Edition: https://createspace.com/5495773

Orlóci, L. 2014. The vegetation process. A holistic study of long-term community energetics in East Beringia. SCADA Publishing, Canada. Online Edition: https://createspace.com/4760258

Orlóci, L. 2013. Quantum analysis of primary succession. The energy structure of a vegetation chronosere in Hawai'i Volcanoes National Park. SCADA Publishing, Canada. Online Edition: https://createspace.com/4452597

Orlóci, L. 2013. Quantum Ecology. Energy structure and its analysis. SCADA Publishing, Canada. Online Edition: https://createspace.com/4406077

Orlóci, L. 2013. On the Energy Structure of Natural vegetation. In search for community governance rules. SCADA Publishing, Canada. Enlarged Online Edition: https://createspace.com/4153484

Orlóci, L. 2012. Self-organisation and Mediated Transience in Plant Communities. SCADA Publishing, Canada. Enlarged Online Edition: https://createspace.com/35 85127

Orlóci, L. 2012. Statistical Ecology. The quantitative exploration of nature to reveal the unexpected. SCADA Publishing, Canada. Online Edition: https://createspace.com/3476529

Orlóci, L. 2012. Statistical multiscaling in dynamic ecology. Probing the long-term vegetation process for patterns of parameter oscillations. SCADA Publishing, Canada. Online Edition: https://createspace.com/3830594

Orlóci. L. 2011. Problem flexible computing in statistical ecology. SCADA Publishing, Canada. Online Edition: https://createspace.com/3574792

Title ID: 7025180   ISBN-13: 978-1544820163   ISBN-10: 54482016X

Find further information at URL https://sites.google.com/site/statisticalecology/
Please send all communications to: lorloci@uwo.ca

Probing simple vegetation data
All rights reserved © 2017 by L. Orlóci & M. Orlóci

Contents

1. Introduction 6
2. Terminology 7
3. Research site 1957 8
   3.1 Preliminaries 8
   3.2 About the flora 10
   3.3 Early work on forest types 10
   3.4 Field data 12
4. Static structures 13
   4.1 Visible structure 13
   4.2 Latent structure 14
   4.3 Example 17
      4.3.1 Vegetation type a complex 17
      4.3.2 Emergent EBE 17
      4.3.3 Catena a virtual sere 18
      4.3.4 Venn diagram components of EBE 19
5. Dynamic structures 20
   5.1 Preliminaries 20
   5.2 Expanding the analytical horizon 22
   5.3 Catena: a proxy succession sere 22
   5.4 Analytical tools 24
      5.4.1 TableCurve 2D 24
      5.4.2 Derivative-calculator 25
      5.4.3 CaptureWizPro 25
   5.5 Example 25
      5.5.1 Numerics 25
      5.5.2 Discussion of exercise 34
      5.5.3 Some questions and a step forward 36
   5.6 Directedness 40
      5.6.1 Preliminaries 40
      5.6.2 Numerics 42
6. The broader context 45
References 47
Index 49
Appendix 53

The objective is to outline concepts and introduce the modus operandi of probing multistate phytosociological data by advanced techniques for information potential regarding complex structural traits in the source vegetation. The analysis of actual data indicates unexpectedly high information potential. But for optimal result, phytosociological gestaltism has to be a part of the sampling design.

1. Introduction

I have had ideas for writing this monograph for quite some time, after re-reading a student essay of mine dating back to my undergraduate years (Orlóci 1958)[1]. But it had to wait until now in need of sufficient time to do it.[2]

The original essay describes stand-level vegetation structure, vegetation types, and successional sere based on a field survey[3] completed in 1957 in Sub-Boreal British Columbia.[4] As I am revisiting the topic, this time my focus is on static and dynamic structural traits, not considered in the 1958 essay.

I presented the barebones of the present monograph in two previous technical reports.[5] At this time, I am amalgamating the two reports and expanding the narrative considerably which takes into account my present conceptual world, and perhaps, sets up a new analytical model for phytosociology.

The 1957 essay's data set and historic account are used after minimal change. The analytical approach is new, designed to probe the data set for intrinsic stand structural traits, which have not offered themselves up for identification in the course of traditional phytosociological synthesis six decades ago. The exercise, as I am unfolding it on these pages, does in fact prove that it was not a wasted time to re-analyse the old data. The results show high information richness of the original, simple phytosociological data set, far beyond the immediate pur-

---

[1] The 1958 essay has been submitted to my undergraduate mentor, Professor Ferenc Tuskó of the Sopron Division in the Faculty of Forestry of the University of British Columbia, in partial fulfillment of requirements for the BSF degree.

[2] Perhaps I waited too long. My most likely and wished for readers – Ferenc Tuskó, Win Arlidge, Art Prochnau, Vladimir J. Krajina, Thomas M.C. Taylor – are no more with us. Even the Research Forest, the survey site of my data set, has been shut down.

[3] I received much help in the field survey from my good friend and fellow Sopron student, György Leskó.

[4] At the time of the field survey, I had in mind a research project to be continued at future dates in an increasingly refined manner. It was not to be. Graduate school - under Professor Vladimir J. Krajina, my mentor in forest ecology at U.B.C. - opened new wistas for me, offering inviting carrier opportunities. I moved on in the direction of quantitative ecology as a NATO Science Fellow in the U.K. under Professor Peter Greig-Smith's guidance at the University College of North Wales.

[5] Forest types and stand structures in V.J. Krajina's sub-boreal spruce forest zone in British Columbia, Part I and Part II. ResearchGate 2016.

pose at the time of its collection. This should draw the attention of fellow researchers, having large phytosociological data sets similar to mine - just as enthusiastically collected, then synthesised by the classical phytosociological methods, but left on the selves for a better day to be revisited. Those data sets may deserve a second look.

What is being presented? The first chapter is a general introduction, the second describes technical terms, and the third presents the conditions as I found them on the survey site in 1957. The fourth chapter presents the method and results of static structural analysis. The structural scalar is energy-based entropy (EBE), an accepted proxy for scaling the potential energy level of vegetation stands. The fifth chapter takes the analysis into the problem area of structural dynamic in the vegetation stand's assembly/disassembly process, in the context of succession theory. These are followed by a chapter that could be read as an introduction a posteriori, left for last because its appreciation requires the experience gained from attentive reading of the main text. References and Appendix of the data set complete the monograph.

2. Terminology

The focus is on two types of objects. One is concrete, such as any *vegetation stand*, and the other is virtual in the manner of the *vegetation type*.

Each stand is described by the *cover estimates* of its component taxa. Cover is estimated on a discrete scale (see in Appendix). Stands are delineated in the vegetation cover as areal units on the ground. A delimited area is referred to as *sample plot*. The *structural traits* in the main title are stand-level traits. An example of such a trait is in the footnote.[6]

---

[6] It is practical to view a stand's energy-based entropy (**EBE**) as a stand-level structural trait, on the basis of which we can write three dimensional model for the structure itself: $E = E_{Phy} + E_{Env} + E_{Rnd}$. Accordingly, stand-level structure is considered as a manifested response to the sum of three independent effects. Quantity E on the left side of the equation and the first two quantities on the right side are measurable. E_{Phy} is proportional to the amount of energy spent by the phylogenetic process to produce the current level of plant taxon richness. The concomitant E_{Env} is proportional to the amount of energy spent in the process of recent environmental mediation in the sorting of plant populations into distinct vegetation stands. The third term is a difference: $E_{Rnd} = E - E_{Phy} - E_{Env}$. Following statistical practice, we assign the unidentified effects responsible for E_{Rnd} to the grab-bag called error generating events.

Homogeneity is assumed within the sample plot in current dimensions of vegetation composition and structure, climatope, edaphotope, and disturbance regime. In all cases, homogeneity implies the random spatial arrangement of stand elements and environmental effects. When the arrangement is a mosaic of patches, homogeneity implies random arrangements within the patches and a random spatial arrangement of the patches within the sample plot. It is not difficult to expect random natural spatial arrangement of plants in environmentally homogeneous sample plots, knowing that the natural stand assembly/disassembly process is responding to chance effects.

The vegetation type's descriptors are averages. We assume homogeneity within the sample plots in order to be justified to assume a Gaussian distribution for the averages, which we are required to do to satisfy the regularity conditions of EBE related analytical techniques. Type descriptions in the 1958 report are based on *phytosociological synthesis* of the stands' record set.

It is intuitively true that a stand's observed compositional state is a passing moment in the ongoing assembly/disassembly process, and so is the processes dependence on environment mediation.[7] The result is a *time series* or *chronosere* of successive plant communities in situ. The product is a directional compositional change in pursuit the momentary attractor state, called *climax*. The attractor is a randomly oscillating target in time. The 1958 essay presents hypothetical pathway of local chronoseres mapped in diagrams. Since the diagrams supporting data set is broader than the one I present in Appendix, I do not consider those aspects in the present monograph. I have the same thing to say about the provisional synsystematic I in the 1958 essay.

3. Research site 1957

3.1 Preliminaries

We now take a virtual visit to the no longer existing Aleza Lake Experimental Station, the site of the 1957 survey, located about 70 kilometres north-east from Prince George on the Upper Frazer road in British Columbia. The station has

---

[7] As a must for paying homage to a historic grate in Ecology, I mentioned the 19th Century naturalist Kerner von Marilaun (1863) who refers to the vegetation process in situ as *community development*. An alternative term is *succession by facilitation*. Kerner describes community development in a manner for which the modern term is a *feedback*. In other words, the vegetation changes its environment and in turn the environment acts back on the vegetation's composition.

been shut down years ago, but a small portion of total area is kept as an ecological reserve. The province wide reserves system is the crowning achievement of Professor Vladimir J. Krajina.

The landscape at the site is dominated by low hills and uplands which form a complex pattern with wetlands. The glacial outwash substrate shares a complex spatial pattern with river alluvia and organic deposits. The climate is Köppen Dfb. The terrain's average elevation is 700 meters above sea level.

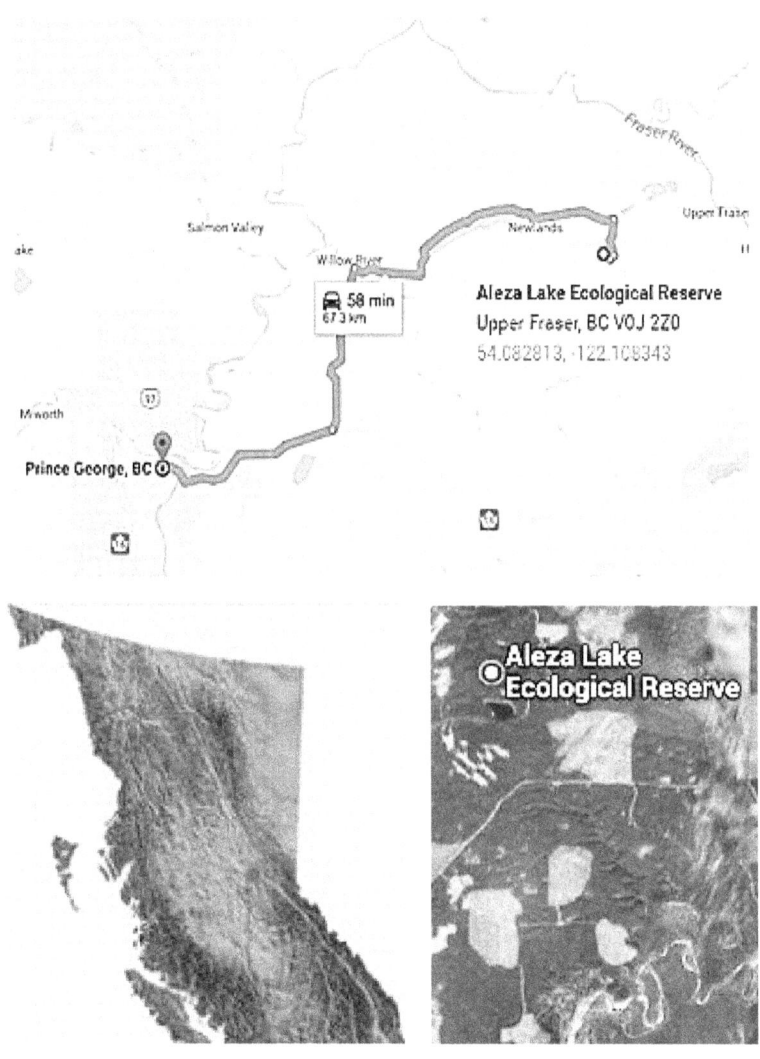

Figure 1. Road map (top) from Prince George to the Aleza Lake survey site. Large dot on map in bottom left locates the site in British Columbia. Some vertical distortion applied. The phytosociological survey covered an approximately 5x8 kilometer rectangular area, including the Ecological Reserve site, all the way down to the Boron River.

3.2 About the flora

My 1958 essay emphasized that the region's topography has no significant barrier for plant distribution. The general climatic effect can be clearly manifested in a floristic continuum. Indeed, the local flora exchanges species with three major biomes in as many different climatic macro zones:

1. The general presence of Douglas-fir and associated species indicate the floristic connection to the Caribou Park Land further to the south on the arid interior plateau, rich in graminoid and semi-shrub species characteristic for the Ponderosa pine savannah.

2. To the north lies the main body of the Sub-Boreal Spruce Forest to which the site's floristic linkage is strong, most typically on deep loam and in muskegs.

3. Out of the south east, species of the Columbia Cedar-Hemlock Forest reach the local sites in numbers on the alluvia laid down by the swift flowing Boron River.

3.3 Early work on forest types

Regarding vegetation studies of historic significance, the 1978 essay mentions Griffith (1926), Kujala (1945), Fraser and Alexander (1949), and Arlidge (1952, 1956). Griffith created the research station's herbarium. I found it poorly kept in 1957. V. Kujala, a research forester on visit from Finland, is apparently the first who had surveyed forest types on the site. His survey has been completed in the summer of 1931, but the publication of his results did not follow until 14 years later. He has named four types based on the leading species in the lesser vegetation:

1. Vaccinium membranaceum
2. Tiarella – Rubus pedatus
3. Tiarella – Fatsia
4. Inpatiens – Circaea – Athyrium

Kujala's classification has no types named from muskegs (peatbogs) or other wetlands.

J.W.C. Arlidge's reports take exception to Kujala's classification, because of his inclusion of Type 1. According to Arlidge, Vaccinium membranaceum becomes

11 | Probing simple vegetation data

dominant under intensive light effect over a wide range of site conditions after logging and fires, and on that basis Type 1 is a one-rotation derivative community. Arlidge recognises six forest types, neither based on short-lived dominance. He names types after leading species in the lesser vegetation:

1. Devil's - club
2. Disporum
3. Sarsaparilla – Oak Fern
4. Bunchberry – Moss
5. Horsetail – Peat moss
6. Black Twinberry - Nettle

The 1958 essay follows Kujala regarding the derivative types. The essay argues that notwithstanding the short longevity of some dominance based vegetation types in situ, on the broader regional scale these types have continuous existence, owing to the regularity of fires and logging. Five of my types are referenced in the data tables of the Appendix. Abbreviated descriptions are extracted from the original text:

Type 1: Black spruce - Sphagnum moss
This type is common on the oldest peat resting on mineral soil at the edge of peatbogs (muskegs in local terminology). Tree height decreases outward from the edge toward the centre where the peat is floating. At its best, on the nutrient rich terrestrial edge, the type-defining Picea mariana (Black spruce) can attain 20-22 metres height. The type is considered an edaphic climax in peat bog succession. The type's indicator species include Ledum groenlandicum, Vaccinium oxicoccus, Kalmia polifolia, and different Sphagna (palustre, rubellum, recurvum).

Type 2: Alder–Struts fern
This type is specific to alluvial soils along creeks and especially on the floodplain of the Boron River. Alnus tenuifolia (Alder) is the dominant tree. It can attain 6-7 m height. The types indicator species include Matteucia struthiopteris, Athyrium filix-femina, Urtica Lyallii, and numerous other nitrophilous herbs.

Type 3: White spruce–Subalpine fir
This type occupies a very wide niche. Picea glauca (White spruce) and Abies lasiocarpa (Subalpine fir) come to their best productivity on deep moist soils. The tallest trees can exceed 40 m height at 80 years of age. Other highly productive species include Picea engelmannii and the sporadic Pseudotsuga mensiesii. The indicator species include Oplopanax horridus, Aralia nudicaulis, Disporum oreganum, Dryopteris disjuncta, Cornus canadensis, Hylocomnium splendens, Rhytidiadelphus loreus, and Entodon shreberi. Patches of Sphagnum squarrosum are common on the humid raw humus. On drier sites, in

southern exposure, Corylus and Vaccinium are abundant. Oplopanax is an indicator of the most productive sites for Picea glauca, and Disporum for Pseudotsuga menziesii.

Type 4: Lodge pole pine–Lichen
This is a one rotation derivative type. It is expected on excessively drained soils following fires and clear cutting in the White spruce–Subalpine fir type. Pinus contorta (Lodgepol pine) shares dominance with Pseudotsuga menziesii, Picea mariana, and Populus tremuloides. Other indicator species include several species of Vaccinium, Spirea, and Viburnum. The ground cover of bryophytes (Rhytidialdelphus, Hylocomnium, Entodon), lichens (Cladonia and Peltigera) are characteristic. Mature trees can top 20 m in height. This type disappears once spruce and fir regaines dominance in the site.

Type 5. Willow – Willowherb
This is a one rotation, derivative of the White spruce–Subalpine fir type on sites laid open by clear cutting or fire on heavy, moist soils. All species of the original type are present. Salix (willow species) and Populus tremuloides (aspen) are well-performing pioneer species. Epilobium (Willowherb), Impatiens, and several fern species attain high abundance. The type disappears once spruce and fir regain dominance over the site.

3.4 Field data

The 1957 vegetation records are reproduced by type in the Appendix. I take from the source data a set of richness and cover totals and sort them in Tables 1a,b by vegetation layer and type.

Tables 1a and b. Total cover (T) and taxon richness (n) stratified by vegetation layer and vegetation type.

	Layer	Cover %					Sample
		Type 1	2	3	4	5	T
1	a	17	1	160	32	10	220
2	b	29	40	219	84	45	417
3	c	33	110	394	102	66	705
4	d	24	0	85	37	0	146
	Sample T	103	151	858	255	121	1488

Table 1b.

	Layer	Richness					Sample
		Type 1	2	3	4	5	n
1	a	4	1	10	7	2	12
2	b	15	10	30	21	16	45
3	c	14	31	55	33	33	85
4	d	5	0	9	8	0	14
	Sample n	38	42	104	69	51	156

Symbols: a-crown canopy, b-shrub layer, c-herb layer, d-ground cover of bryophytes and lichens; T total cover (by area %); n total number of species contributing to T.

Taking note that cover values are scored on a discrete scale. Therefore, we have to expect the sum of cover values to top100 % within a sample plot. I discuss structure in more detail under two headings in the sequel. One is the static structure which is momentary, usually referred to as time-static. The other is dynamically unfolding in time.

4. Static structures

4.1 Visible structure

The number of layers in the stand, or equivalently plant growth forms (trees, shrubs, herbs, cryptogams) as I use this term, the number of species, cover totals, and the distribution of these among layers or on the ground, allows me to speak about stand structure as a visible object. This type of structure can be seen, directly measured and described. The graphs in Figure 2 depict a compound stand structure based on cover and richness totals over taxa by layer and vegetation type.

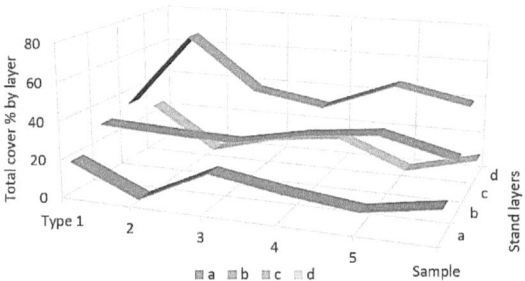

Figure 2a. The T graphs.

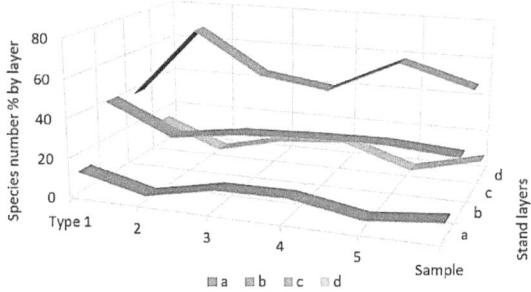

Figure 2b. The n graphs

Figure 2. Graphs of T and n values over type and stand layer based on data taken from Table 1.

Looking at the graphs in Figure 2a, the arrangement by layer is best seen on the west face of the virtual box. This shows the herb layer (c) with highest graph in each type, the shrub layer (b) comes next, followed by the cryptogam layer (d). The tree layer (a) is last. The same pattern is observed for n in Figure 2b.

It is obvious from examination of the eigenvalues and eigenvectors in Table 2 that cover % and richness % capture statistically identical structures. The specific eigenvalues recover the covariance structure with 85 and 95 percent efficiency. The correlation of the two eigenvectors rounds up to 1. In these terms, taxon cover and taxon richness while generically different, in essence can be considered as proxy structural traits in my simple phytosociological data. To put the local findings into perspective, I point to a strong binary effect, meaning that the data set (Appendix) is saturated with zeros and ones. This is normal for phytosociological data from preferential sampling.

Table 2. Eigenanalysis of the layer by type data sets (Table 1).

Source	Eigenvalue	Efficiency %	Eigenvectors					Correlation
Cover%	1495.1	89	-857.9	1061.5	-203.5	-406.9	406.8	1.0
Richness%	1127.7	95	-712.0	722.8	-101.8	-292.0	383.1	

4.2 Latent structure

This structure is real, but it reveals itself only in the analytical space. An example has already been given earlier in the regarding stand level *potential energy structure*. For this, I used as my proxy scalar Max Planck's *energy-based entropy* (EBE). I did it in connection with the resonator complex model of the stand. I describe the details on the pages of Statistical Quantum[8] Analysis (Orlóci 2017).

Table 1 is an example for at least two variants of the resonator complex model. One is for vegetation types and the other for stand layers. The resonators are plant taxa. Depending on the case, the identity of the complex and its parameters T and n will change, but the functional form of EBE (abbreviated by E) remains the same:

$$E = -\ln P = \ln \frac{(T+n-1)!}{T!(n-1)!} \approx \ln \frac{(T+n)^{T+n}}{T^T n^n}.$$

---

[8] The term 'Quantum' honours the historic fact that the analysis involves Max Planck's (1901) energy-based entropy function and idea of a resonator complex model.

15 | Probing simple vegetation data

When we write the equation E for a given vegetation type (Table 1), parameter T and n refer to that type. Written for a given layer, T and n refer to that layer.[9] Regarding parameter P, it defines the commonness of the resonator complex.[10] The basics are reviewed:

a. $E = nH = (T+n) \ln (T+n) - T \ln T - n \ln n$

This is the 'working' equation for EBE, and in proportional terms for the potential energy level in the complex. Clearly, E can be interpreted as a high level, holistic ecological entropy parameter. Note further that symbol ln indicates Natural logarithm. This is why we say that E is measured in *natural units* or in *nats*.[11]

b. $H = \dfrac{nH}{n}$

E is dependent on n, H is not. Therefore, H is comparable between cases. E is unique to the case.

c. $P = e^H$

This is the probability of an H, exactly as extreme as the observed H occurring by chance alone. On this basis, we can test the statistical significance of any H or its partitions (see EBE model in Chapter 2).

d. $w = 1 - P^2 - (1-P^2)$

This is a squared probability, proportional to EBE instability in the complex's The value of w ranges between 0 (complete stability) and 0.5 (complete instability). This is easy to see if 1 or 0.5 is substituted for P. How do we interpret w? There is more than one way to do this. One example: when w increases, the chance of the stand's energy state flipping by pure chance into one of its **possible alternative** states increases. In other word, the chances of the stand's composi-

---

[9] According to the original definition (Planck 1901): T stands for the total number of energy units normally distributed among n resonators. In our usage one cover unit is equal to 1 energy unit.

[10] $P = \dfrac{1}{C}$ and $C = \dfrac{(n+T-1)!}{(n-1)!T!} \approx \dfrac{(n+T)^{n+T}}{n^n T^T}$. In other words, P is the probability of drawing a complex at random from the family of C distinct complexes with common resonator number n and total energy unit count T. The complexes are equiprobable.

[11] Energy cannot be measured directly. Proxy functions measure the energy's manifestations. Max Planck offered EBE. He reasoned that the complex's probability, a complement of its uniqueness, is a proxy scalar for the complex's energy level in the manner of - ln P. Proxy scalars are proportional to the unknown energy quantity.

tion changed by chance into compositional state increases. The number of possible states is $\frac{1}{P}$. The choice may fall with equal probability on any of the C-1 states with $\frac{1}{C}$ probabiltiy.

e. $\omega = \sqrt{2w}$

This is a probability associated with w. The ω parameter reappears in two other parameters, $^-m\omega$ and $m\omega$.

f. $^-m\omega = -\ln(1-\omega)$

We refer to $^-m\omega$ as the *unit instability moment*. It is measured in nats. This is energy-based entropy, the strength of instability, or equivalently, the unit linear moment forcing the stand's energy structure to flip into another random state. All possible states, except the state actually observed, are called *ghost states*.

g. $m\omega = -\ln(\omega)$

This is the *unit stability moment* (nats).[12]

Considering Table 1, it is intuitive that a layered structure to become the same as the observed, requires a specific amount of energy. It is intuitive too that a structure defined is a structure measured, therefore, the structure revealed is specific to the scaler function which is used to measure it. This is the same as saying that when I am using the EBE function to measure vegetation structure, I will end up, by virtue of using a proxy function, with a potential energy structure measured in natural units. The idea that EBE is a valid scalar of the vegetation's energy structure is linked with the idea that the vegetation stand can be modelled as a *resonator complex*. The use of EBE need not be restricted to the nano world of our reality.

For us, any homogeneous vegetation stand, or collection of stands, is a *resonator complex*. Considering that Max Planck's EBE is a function of the probability P of the resonator complex, when P is given EBE is defined. Further, EBE is proxy function to estimate the potential energy level of the vegetation complex. In this case, we pair up one performance unit (cover unit in the example) with one potential energy unit. Note that certain regularity conditions apply:

---

[12] Another kind of moment is the standard deviation, 2nd central moment, having to do with variation about the mean.

17 | Probing simple vegetation data

(1) The distribution of the energy units among the resonators follows a Normal distribution. For us, this implies a random mechanism at work in convolution with effects issuing from evolution and environmental mediation.

(2) Probability P is known.

(3) The homogeneity assumption (already discussed) holds true for the complex.

4.3 Example

4.3.1 Vegetation type a complex

Table 4 contains a complete set of results. Note, T and n are vegetation type totals taken Table 4. Everything else is calculated according to he equations already given.

Table 4. Statistical analysis of types-based E. Since E is proportional to potential energy, the results allow approximation of the potential energy level of the types and evaluation of the EBE structure's stability.

Parameters for Layer (Table 1b)	Type 1	2	3	4	5	All types	Excluding Type 5
T	103	151	858	255	121	1488	1367
n	38	42	104	69	51	*156	**105
		101.10	329.52	167.78	104.55		
nH	82.170	8	5	5	5	515.738	378.407
H	2.162	2.407	3.169	2.432	2.050	3.306	3.604
P	0.115	0.090	0.042	0.088	0.129	0.037	0.027
1-P	0.885	0.910	0.958	0.912	0.871	0.963	0.973
P^2	0.013	0.008	0.002	0.008	0.017	0.001	0.001
$(1-P)^2$	0.783	0.828	0.918	0.832	0.759	0.928	0.946
$w=1-P^2-(1-P^2)$	0.204	0.164	0.081	0.160	0.224	0.071	0.053
$\omega=(2w)^{0.5}$	0.638	0.573	0.401	0.566	0.670	0.376	0.325
$m\omega=-\ln(\omega)$	0.449	0.558	0.913	0.569	0.401	0.979	1.123
$m\omega=-\ln(1-OM)$	1.017	0.850	0.513	0.835	1.108	0.471	0.394

*Total number of taxa in the pooled sample of 5 types. **Total number of taxa in the pooled sample of types 1 to 4.

The five types can be ordered on any of the parameters. The instability parameter $w=1-P^2-(1-P^2)$ is a good example. Type stability order by 0.5-w is 3,4,2,1,5. Type 3 has highest stability. Ecologist consider Type 3 as the climatic climax. Type 5 is the most labile. It is in the recovery process.

4.3.2 Emergent EBE

The *emergent energy-based entropy* or *ghost energy-based entropy* (gnH in table below) comes about by enlarging the complex. In the example, I simulate this with the 5 types in two steps. As the first step, I pool Types 1, 2, 3, 4 and calculate E. In the second step, I add Type 5 to the group of Types 1, 2, 3, and 4. This is what we get:

nH(1 to 5)	nH(1 to 4)	dnH	nH(5)	*gnH
515.7376	378.4065	137.3311	104.5553	32.7758

* Ghost EBE

The last cell has the size of the emergent or ghost EBE, generated when group of Types 1, 2, 3, and 4 is enlarge by adding to it Type 5. The increase in EBE is 32.78 nats or 100 gnH/nH(1 to 5) = 6.4%. This may remind the reader of the sagely dictum: the whole can be greater than the sum of its parts.

4.3.3 Catena a virtual sere

Figure 3 shows an arrangement of Types 2, 3 and 4 on a virtual sere in order of their decreasing average moisture regime. Type 1 (not shown) is included in the calculations. The base data are found in Table 1 and the results of the analysis in Table 5. The objective is to isolate the phylogenetic effect (E_{Phy}), linked to functional plant type, enabling layering in the stands, from the environmental effect (E_{Env}) associated with soil moisture regime.

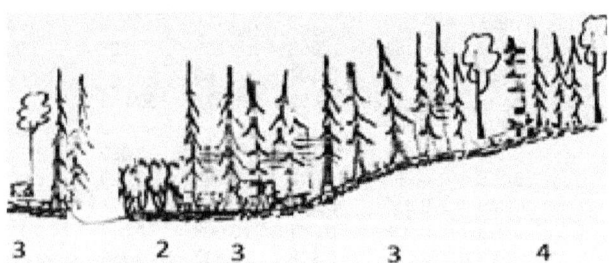

Figure 3. A virtual catena cutting through a stream and hillside. Numerals identify the types as described. Types 1 and 5 are not shown. See the main text for details.

Table 5.

Types	T	n	nH	dnH	dn	dH	P	H	P
1	103	38	82.170	82.170	38.000	**2.162**	**0.115**	2.162	0.115
1+2	254	74	175.125	92.955	36.000	**2.582**	**0.076**	2.367	0.094
1+2+3	1112	134	425.322	250.197	60.000	**4.170**	**0.015**	3.174	0.042
1+2+3+4	1367	151	491.716	66.394	17.000	**3.906**	**0.020**	3.256	0.039
				491.716	151.000				

The symbols' identity has been defined earlier in the text. Column dH and the next column P are of interest at this point. These have corresponding graphs in Figure 4. Interesting to observe that by crossing into the mesic zone (Type 3) from Type 2 on the virtual catena. Taking 0.05 as the critical probability, the maximum reached is statistically significant.

19 | Probing simple vegetation data

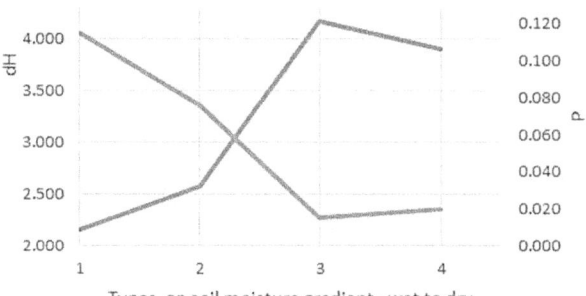

Figure 4. dH minimum in Type 1, maximum in Type 3.

4.3.4 Venn diagram components of EBE

The E quantity can be partitioned into components based on the Venn diagram's logic. I refer to earlier papers (Orlóci 2006, 2015) for examples and references. There are main effects, random effects, interaction (mutuality) effects, and total effects. All effects are parts in additive sequences on the EBE axis. We make the partitions specific to the effect of phylogeny (functional type or taxon richness), environmental mediation (moisture gradient), and chance events. Table 6 contains the results for layering and in Table 7 for taxa.

Table 6.

	T	n	nH	%	nH$_{Layer+Emergence}$	H	P
Layering	1367	4	27.3422	31.33	59.9157	6.6573	0.0013
Environment	1367	4	27.3422	31.33	59.9157	6.6573	0.0013
Emergent*	1367	**5	32.5735	37.33		6.5147	0.0015
Joint	1367	16	87.2578	100.00		5.4536	0.0043
Total					119.8313		

* Also known as interaction, **n value by iteration:

T	n	nH
1367	4	27.3422
1367	**5	33.0638
1367	6	38.5848

On inspection, we find in Table 6 the emergent effect topping 37% in absolute terms, but the relative values (column H) are not all that different. The probabilities indicate highly significance levels of EBE.

Table 7.

Components	T	n	nH	%	nH$_{Rich+Emer}$	H	P
Richness	1367	151	491.7158	39.11	1188.0257	3.2564	0.0385
Environment	1367	4	27.3422	2.17	723.6520	6.8355	0.0011
Emergent*	1367	**250	696.3099	58.36		2.7852	0.0617
Joint	1367	604	1214.5802	100.00			
Total					1911.6778		

* Also known as interaction, **n value by iteration:

T	n	nH
1367	249	694.4413
1367	**250	696.3099
1367	251	698.1751

Table 7 is constructed for 151 taxa in the first 4 types. The results indicate that the environmental effect is negligible in absolute terms, but not so in terms of H.

5. Dynamic structures

5.1 Preliminaries

The discussion of dynamic structures should begin with an often applied assumption that structural change of any kind in the vegetation's composition in situ over time has a proxy model in space-for-time substitution for which our vehicle in the example is the virtual catena. This may sound an extravagant assumption, but really it is not. Nor it is a novel one. In fact, the idea is as old as succession theory itself.

I give two examples. The long-term vegetation dynamic we find mapped by the distribution of palynomorph taxa within the vertical core of a peatbog has parallel in the current latitudinal zonation of the vegetation and climates on once glaciated terrain in the cool temperate cold regions of the Northern Hemisphere. In a similar way, changes in vegetation composition in situ at any point of a flood plain's quickly aggrading terrain, has parallel in compositional changes seen on the elevation gradient on the inner bank of a meander. The principal ordinal of the assumption is the realization that space and time is interchangeable proxy dimensions within the natural succession process; therefore, the vegetation we see around us is a collection of stands that reflect back on compositional states past and forward on states in the succession process.

Kerner von Marilaun (1863) was the first to ask a scientific question about the vegetation process. Namely, what drive compositional change (Kerner's vegetation development) through time in the natural environment, the kind we refer to in today's parlance as *vegetation succession*? In other words, what causes transformation of one vegetation type into another in the same site? Kerner's answer in his 1863 classic is profoundly scientific in its reasoning and conclusion, based on field observation of spatial vegetation pattern an active alluvial talus. His method allowed Kerner to connec space to time, and to conclude that pattern observed in space has a dual in time. He deduced that natural communities develop in situ. This process is simultaneously performing a dual function in the manner of compositional assembly and disassembly in the plant community. Kerner found a reason for this in the functioning of the plant community assembly which causes change in the site, progressively less favorable for the species inside more favourable for new comers from outside. Today's technical term for Kernerian community development is *succession by facilitation*.

Kerner's discovery heralded paradigm change in the naturalist's world from a static and largely aesthetic view of the vegetation to the scientific view which sees the vegetation as a dynamic, ever changing medium. Kerner's work is about a high-level process in Nature, not any less important in the rise of modern Natural Science than the Darwin-Wallis theory of species evolution, the Mendel's theory of particle based inheritance, or Dokuchaev's theory of pedogenesis in vegetated sola.

In this chapter we take an aim at an empirically derived integral function $Y=f(X)$ and the associated 1st and 2nd order differential equations in the study of assembly/disassembly dynamic, call it succession, based on our virtual catena of the types. The symbol X stands for position "no" on the catena, and Y is an alternative symbol for nH. In mentioning the analytical tasks, to be completed in the order of finding the integral function first and then turning to the study of its properties, goes head on against the Lotka-Volterra[13] bandwagon's direction. I do recognize the high complexity of the medium I am working with, which keeps me from starting with rates expressed in the manner of the Lotka-Volterra *logistic functions*.[14] I dich that approach. I do this since from its direction I cannot construct differential equations for a complex process. Instead, I begin with the integral equation itself, which I construct from observational data - when I have no theory to guide me otherwise.

My approach is simple and practical. First, I extract the integral function of the observed natural process from observational data, and then find the functions extreme points. My tools are the differential equations. This approach paves the way for the application of higher mathematics in phytosociology. The phytosociologist need not be a mathematician to get involved with this. Let us face it,

---

[13] Refer to https://en.wikipedia.org/wiki/Lotka%E2%80%93Volterra_equations for informative details.

[14] A.J. Lotka and V. Volterra introduced simple differential equations, called logistic functions, in biology. It is worth noting that the integral for these is always an exponential. But really, who has ever seen a complex process in the real world, such as the plant community's succession, to have exponential behaviour? Is any theory telling us that the succession process has to be unfolding on an exponential path? Another problem with starting with rates, when the objective is to generate the integral model of a complex system, is the same as with the reductionists' attempt to build models for complex systems, starting with the behaviour of the system's very basic individual components, as if they had individual existence independent from one another. Such models always bound to l limp, or be outright useless, since that way the all-important system property of emergent behaviour is not detectable. Big science discovered this golden rule decades ago (Gleick 1987), and shifted toward holism in its approach to understanding complex system behaviour. In such an approach intuition and gestalt play an important role.

R.A. Fisher and his contemporaries saw to it that the practitioner does not have to be a probability theorist to apply statistical technics. Why should then the practitioner be required to be a theoretical mathematician to make use of 'higher' mathematics?

5.2 Expanding the analytical horizon

Considering the experience of phytosociologists at large in the Fisherian statistical dialect (Orlóci 1993, 2001, 2016), specifically in analytical structures created by coupling the system of moments - particularly the central moments, such as the sums of squares and cross products - with probability theory, I use the opportunity to introduce the reader to another statistical dialect which can broadens the horizon of statistical analysis in ecology. Its central structures and test procedures are constructed for EBE.

I have considered statistical work with EBE. In this section I consider incorporation of differential equations into the study scenario simply and effectively, and of course, without the usual pomposity. The practitioner should find that differential equations constructed by downloadable applications and used intuitively are no more difficult to use and interpret for most users than the the difficulty they face with application of Fisherian statistics.

5.3 Catena: a proxy succession sere

The term has meaning as a linkage of elements by occurrence in space/time, such as vegetation stands or types, ordered by some strong environmental effect that mediates facilitation and causes succession. The linkage criterion for the present purposes is soil moisture regime.

Phytosociologists often use serial spatial arrangements, real or virtual, such as in Figure 3, as proxy model for the temporal sere of succession. The present catena model attributes facilitation to changing soil moisture in the site, which in turn facilitates changes in the plant community's composition. In this way, the plant community facilitates its own replacement in time. In this sense, there is Kernerian vegetation development, not in a genetically definable way, but in compositional terms mediated by soil moisture.

In the virtual model of Figure 3, the catena runs from wetland to hill top on a relic sand dune. The question that comes to mind is one that Kerner asked in 1863: what kind of process can power plant community development in situ? What does the process do to stand composition under the catena model? In technical parlance, what and where are extreme points on an integral function, particularly the infection points where one forest type gives way for the next? Further, are these points associate with extreme points on the integral function of soil moisture?

23 | Probing simple vegetation data

It is implied in what has been discussed so far that I am interested in the dynamics of the EBE structure in the plant community. Everything I do in the sequel is based on the first 34 relevés of the 37 tabulated in the Appendix. The one rotation derivative type (Type 5) is excluded. Table 8 displays the T, n, nH and H values.

Table 8. Numerical description of the 34 stands, ordered by to soil moisture from wetland (no 1) to well-drained soils on hill tops (no 34).

Stand no	T	n	nH	H
1	27	19	31.1856	1.6413
2	29	21	34.0146	1.6197
3	17	13	20.5270	1.5790
4	30	18	31.7550	1.7642
5	24	17	27.8185	1.6364
6	36	19	35.4523	1.8659
7	34	24	39.3361	1.6390
8	35	26	41.6156	1.6006
9	22	14	24.0569	1.7184
10	53	42	65.2107	1.5526
11	40	31	48.6415	1.5691
12	47	37	57.6277	1.5575
13	51	37	59.8786	1.6183
14	45	34	53.9903	1.5880
15	43	34	52.8452	1.5543
16	55	38	62.9002	1.6553
17	38	29	45.8346	1.5805
18	37	32	47.6458	1.4889
19	52	45	66.9825	1.4885
20	41	32	50.0435	1.5639
21	37	27	43.5770	1.6140
22	47	32	53.3259	1.6664
23	48	34	55.6371	1.6364
24	33	25	39.6490	1.5860
25	38	31	47.4715	1.5313
26	29	22	34.8686	1.5849
27	43	32	51.1765	1.5993
28	39	32	48.8678	1.5271
29	42	32	50.6151	1.5817
30	54	46	68.9944	1.4999
31	51	41	63.2250	1.5421
32	54	47	69.7651	1.4844

33	50	45	65.7173	1.4604
34	51	43	64.8150	1.5073

What next? The first step is to find a shape function (the integral function) which captures the nH series in meaningful terms. "Meaningful" to me, and I suppose to most practitioners in phytosociology, is definitely much more than slavish use of some residual-based error condition. I add, since the shape function will be fitted to the data in a regression analysis, a choice ruled by statistical errors alone, based on residuals, cannot be guaranteed to be an ecologically meaningful choice. Equally as important, the shape function's meaningfulness. The logic in this is that statistical errors are not unique. They can be minimized almost to any extent by changing the order of the polynomial from which the residuals are measured as deviations. To express this in another way, I put two question for further consideration:

1. What is the specific form of the shape integral $f(X)$, which provides meaningful description of the trajectory of nH regarding the soil moisture effect our proxy criterion for time in the aggrading land type modelled by moving upward on Aleza Lake's virtual sand hill (Figure 3)?

2. What are the cardinal properties of $f(X)$ to use as criteria of ecological meaningfulness?

So, what are the tools I must have to tackle these questions?

5.4 Analytical tools

All are rather user-friendly application programs, downloadable from the Internet free or for a modest price. Beware! Look for the authenticated free version, and make a habit to pay a small donation when requested.

5.4.1 TableCurve 2D

This is a venerable piece of application created by brilliant people decades ago. I enquired a copy in the early 1990s. The same version has been running seamlessly on all successive versions of windows on my PCs. TableCurve gives access to over 8000 functions from which I can choose the 'ideal' shape function $f(X)$. The regression fit is automatic, but the choices are the user's responsibility. The application parametrizes $f(X)$ and perform many other things on demand. All results are presented in completely labelled and annotated graphs and formatted data tables of the user's choice. TableCurve 2D is a free application. Should you decide to download TableCurve, do not miss the donation.

5.4.2 Derivative-calculator

This fine marvellous piece of software can be accessed at http://www.derivative-calculator.net/ and run on line. All operations are symbolic. An example. Suppose, the function selected for f(X) is #7903,

$$f(x) = \frac{ex^2 + cx + a}{dx^2 + bx + 1}$$

Copy the operational part (on righ) onto the clipboard, and paste it into the designated box on the screen of the derivative-calculator. Select the option for what you want to calculate and find it created in another large box on the same screen. If you select the "1st derivative", you get the 1st differential equation

$$\frac{d}{dx}f(x) = \frac{-(cdx^2 - ebx^2 + 2adx - 2ex - c + ab)}{(dx^2 + bx + 1)^2}$$

Get the numerical values of a, b, c, d, e from the output of TableCurve. The same values which parameterize $f(X)$ are parameterising $\frac{d}{dx}f(X)$.

5.4.3 CaptureWizPro

For me, this application is simply wonderful. It is making a breeze for me to transport graphs or anything else from the screen into the manuscript - and leave all application still running. CaptureWizPro is worth every penny of its modest price.

5.5 Example

5.5.1 Numerics

How to go about the analysis for finding the shape function and the associated 1st and 2nd differential equation for the data in Table 1? Follow the steps:

1. I assume Table 1 is on a spreadsheet:

A	B	C	D	E
no	T	n	nH	H
1	27	19	31.1856	1.6413
2	29	21	34.0146	1.6197
3	17	13	20.527	1.579
4	30	18	31.755	1.7642
5	24	17	27.8185	1.6364
6	36	19	35.4523	1.8659
7	34	24	39.3361	1.639
...	...	...	...	...
32	54	47	69.7651	1.4844
33	50	45	65.7173	1.4604
34	51	43	64.815	1.5073

Copy the entire data as a block from first cell (no) to last cell (1.5073) to the clipboard.

2. Start processing by clicking on the TableCurve icon on the screen. A new screen comes up which should look similar to this:

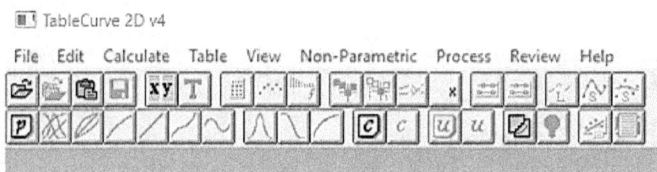

3. Click on the file menu and select "Import clipboard". A box should appear on your screen:

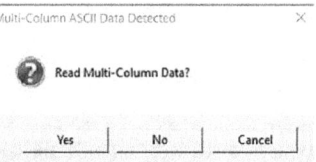

Click "Yes" to read multicolumn data.

4. A new menu comes into view:

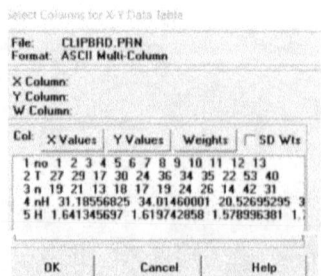

For "X Column" click on 1 under "Col:". For "Y Column" click on 4 under "Col:".

27 | Probing simple vegetation data

5. A revised menu appears on the screen. Click OK. This brings up a new menu. Revise the centre block to read as below:

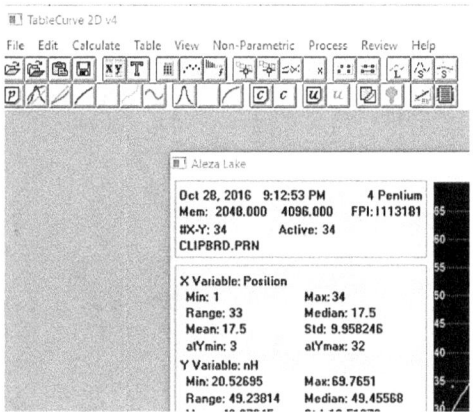

Click OK.

6. The next menu appears:

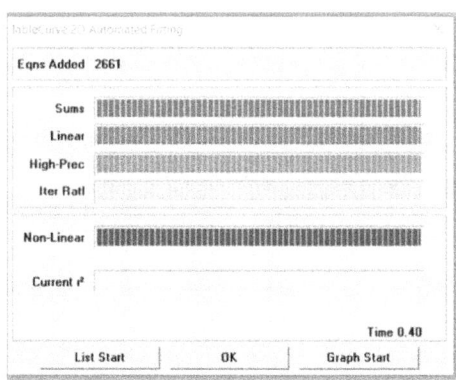

In the second row of icons on top click the second ikon.

7. A new menu appears:

When ready, click "List Start".

8. A list appears on your screen:

■) 2647 Equations [Rank, r², FP, Eq#, Eqn]

File	Edit	List	Filter	Sort		
1	0.7676500018		41	6870	Chebyshev=Std Polynomial Order 20
2	0.7676499982		85	6820	Chebyshev Polynomial Order 20
3	0.7676499548		41	6070	High Precision Polynomial Order 20
4	0.7661074914		147	6850	Fourier Series Polynomial 10x2
5	0.7654250157		91	6810	Chebyshev Polynomial Order 10

Click File 1. The application puts a graph on your screen.

9. Click on "Search" on the side of the graph. A table appears:

Enter 7903 and click OK.[15] This is part of what comes up next on the screen:

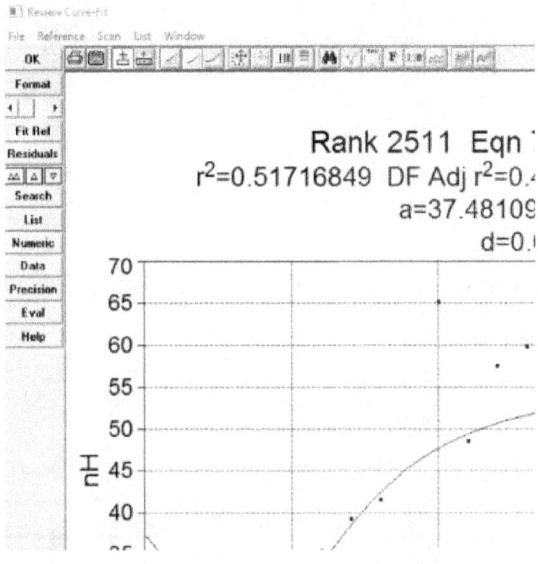

---

[15] I used Equation 7903 as one of my shape functions in the monograph "Statistical quantum ecology. Essays on the resonator complex model of the vegetation stand" (https://createspace.com/ 6509504). The equation has high ecological relevance in whole, and also the numerator and denominator separately.

29 | Probing simple vegetation data

The full screen shows the graph of the function

$$f(x) = \frac{ex^2 + cx + a}{dx^2 + bx + 1}$$

10. Select Format and click the third option in

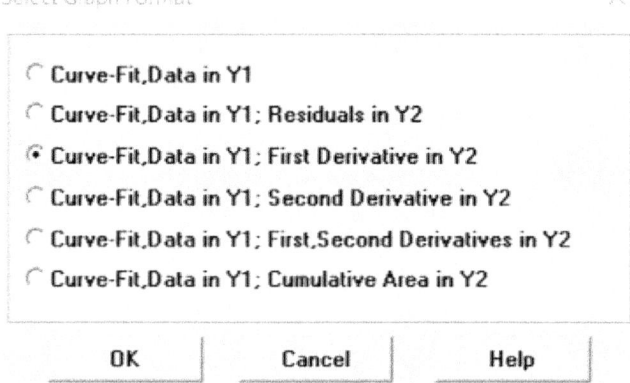

Click OK.

11. A new graph appears:

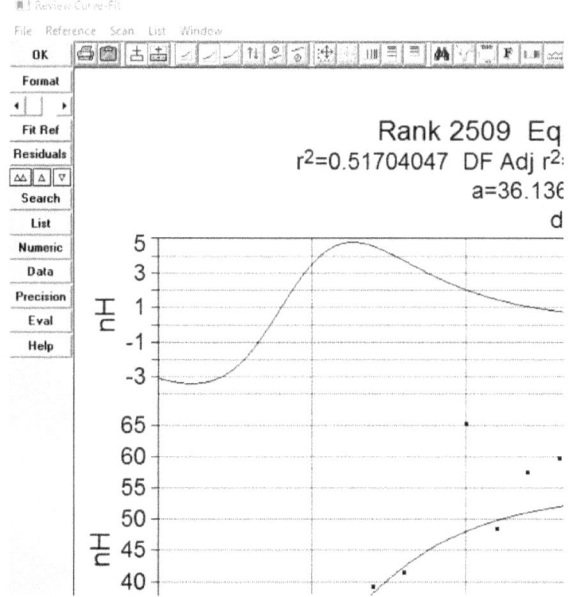

12. Press the 6th icon from right in the full graph's heading and revise the menu's contents as needed to read:

Click OK. The revised graph appears:

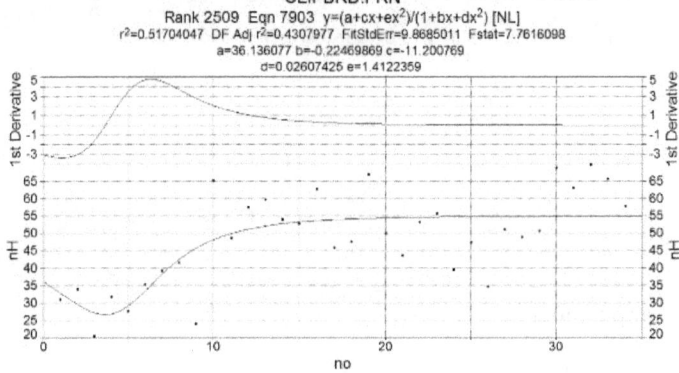

Important: the 1st derivative is scaled in TableCurve as a tangent. It can be negative. Ignore the sign to read.

Pick up the revised screen graph by CaptureWizPro and paste it into your manuscript. Do not close the TableCurve menu.

13. Click Format again (see 2nd graph under step 9) and select 2nd derivative (step 10). The new graphs should look like these:

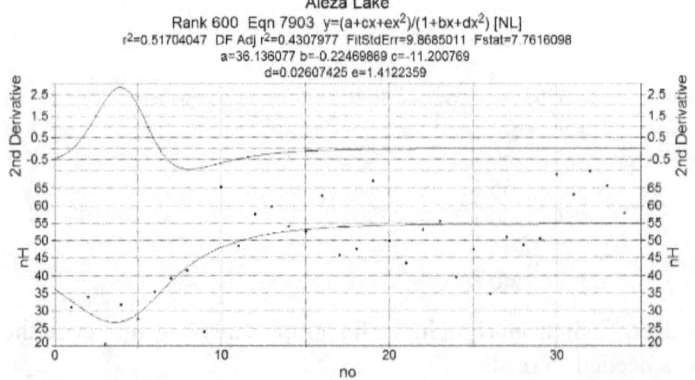

31 | Probing simple vegetation data

Paste this graph too into your manuscript. Do not close the graph in Table Curve yet.

So far we have for TableCurve's equation 7903:

$$f(x) = \frac{ex^2 + cx + a}{dx^2 + bx + 1}$$

$$\frac{d}{dx}f(x) = \frac{-(cdx^2 - ebx^2 + 2adx - 2ex - c + ab)}{(dx^2 + bx + 1)^2}$$

$$\frac{d^2}{d^2x}f(x) = \frac{2\left(cd^2x^3 - ebdx^3 + 3ad^2x^2 - 3edx^2 - 3cdx + 3abdx - ad - bc + ab^2 + e\right)}{(dx^2 + bx + 1)^3}$$

These are equivalent to the Derivative-calculator versions:

YOUR INPUT
$$f(x) =$$

$$\frac{ex^2 + cx + a}{dx^2 + bx + 1}$$

Note: Your input has been rewritten/simplified.

Simplify Roots/zeros

FIRST DERIVATIVE:
$$\frac{d}{dx}[f(x)] = f'(x) =$$

$$\frac{2ex + c}{dx^2 + bx + 1} - \frac{(2dx + b)(ex^2 + cx + a)}{(dx^2 + bx + 1)^2}$$

Simplify:

$$\frac{(cd - eb)x^2 + (2ad - 2e)x - c + ab}{(dx^2 + bx + 1)^2}$$

Simplify Show steps Roots/zeros

SECOND DERIVATIVE:
$$\frac{d^2}{dx^2}[f(x)] = f''(x) =$$

$$\frac{2(2dx+b)\left((cd-eb)x^2+(2ad-2e)x-c+ab\right)}{(dx^2+bx+1)^3} - \frac{2(cd-eb)x+2ad-2e}{(dx^2+bx+1)^2}$$

Simplify:

$$\frac{2\left((cd^2-ebd)x^3+(3ad^2-3ed)x^2+(3ab-3c)dx-ad-bc+ab^2+e\right)}{(dx^2+bx+1)^3}$$

Simplify | Show steps | Roots/zeros

14. Click Format once more in TableCurve and select residuals. You should end up with these:

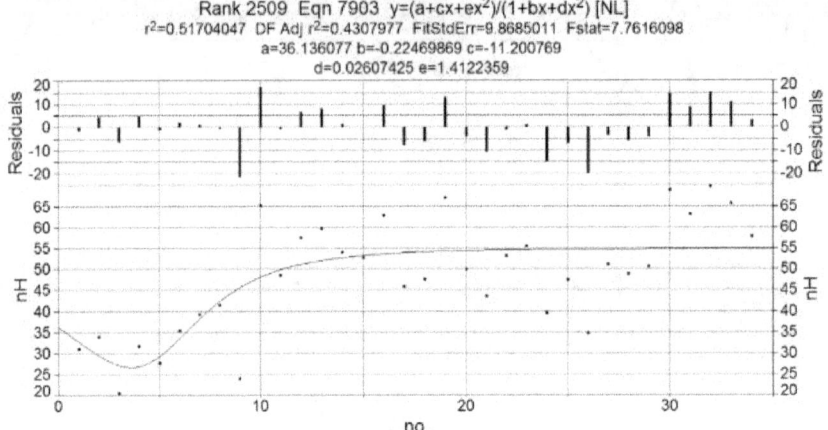

Paste this into your manuscript.

15. Start at the beginning, but at this time select the H column in Table 1 as your Y variable. Select equation 8004 from TableCurves list:

$$f(x) = \frac{b}{\frac{(x-c)^2}{d^2}+1} + a$$

This has 1st and 2nd derivatives, copied directly from the application's page,

$$\frac{d}{dx}f(x) = \frac{-2bd^2 \cdot (x-c)}{\left((x-c)^2+d^2\right)^2}$$

$$\frac{d^2}{d^2x}f(x) = \frac{2bd^2 \cdot (3x^2-6cx-d^2+3c^2)}{\left(x^2-2cx+d^2+c^2\right)^3}$$

TableCurve hands us the following graphs:

33 | Probing simple vegetation data

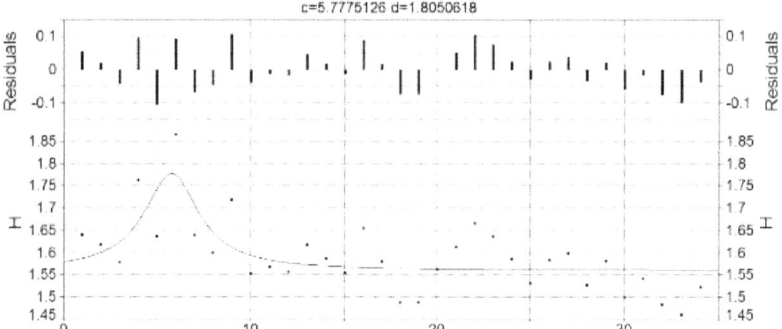

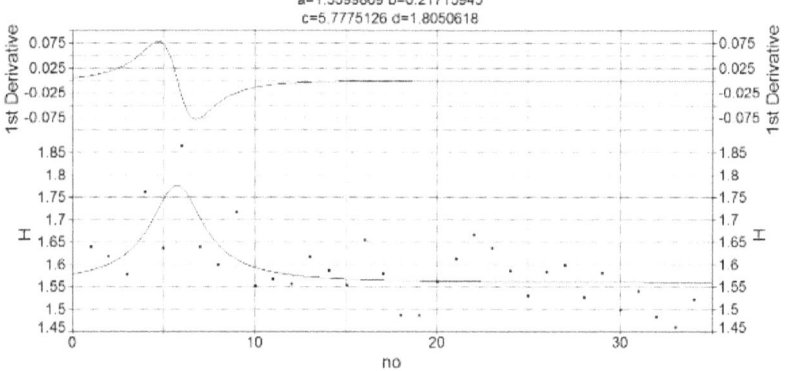

Important: the 1st derivative is scaled in TableCurve as a tangent. It can be negative. Ignore the sign to read velocity.

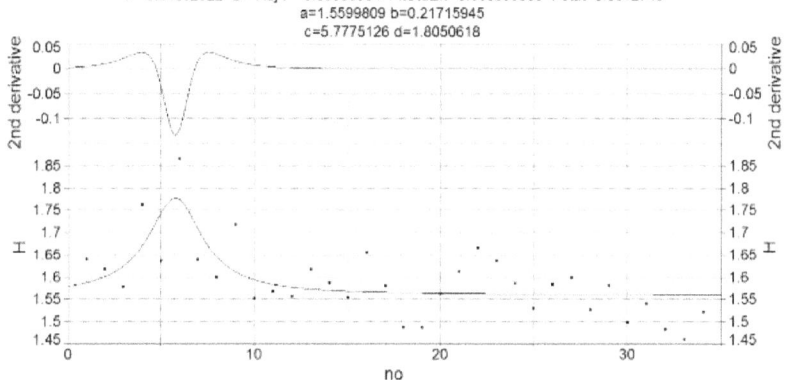

5.5.2 Discussion of exercise

Why did I choose Equations 7903 and 8004 for shape functions? The answer is in my reasoning. I begin with the interpretation of the catena and clarification of the objectives I have set regarding the catena:

a. The catena is my virtual object. It captures a time series in situ. It is an arrangement of the sample plots from different sites according to their soil moisture regime from wet to arid through several grades of mesic. At the low point, there are the wet lands. From there the virtual landscape rises onto the slope of a fossil sand dune, then continues up to some inflexion point where the concave shape of the hill side gives way to a convex upper slope, and further up to the exposed hill top. I expect structural changes in the vegetation in progression from the first to the last, some change in species richness and in productivity. I expect the rate of change to be intensive where the wetland meets the concave lower slope, and albeit not as dramatic but still substantial where the shape of the slope changes from concave to convex. Finally, where the hilltop reached, the compositional richness increase and productivity declines.

b. I have chosen shape functions $f(X)$ that are simple, yet they have the power to signal intense change in nH in c-ordination with catena model. I refer to the points of intense change as *extreme points*. I use derivatives, in actual fact the graphs of differential equations, to identify extreme points such as maxima and minima, or even inflexions where a concave section of the graph goes into convex, or in the reverse. Note that the graph of the 1st derivative measures the velocity of change in nH. Note further, TableCurve presents the 1st derivative as a tangent value, zero when the graph of $f(X)$ is flat, positive going up, and negative going down. For velocity, the negative sign is ignored. The graph of the 2nd derivative fixes acceleration (positive value) or deceleration (negative value) at any given point.

c. Read the following graphs, according to the explanations:

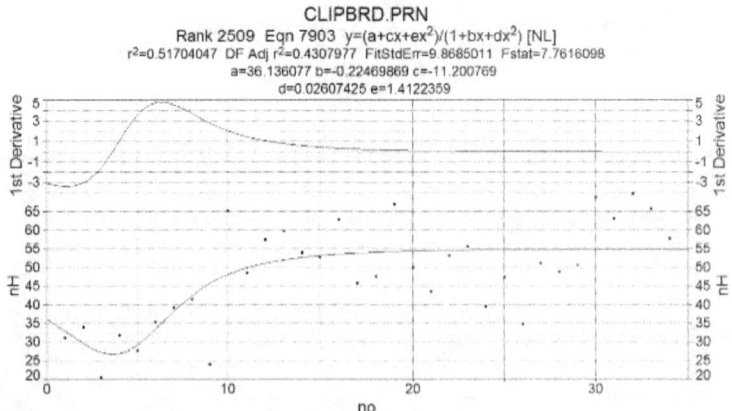

Begin with the 1st derivative. The first extreme point is a minimum, corresponding to a weak inflexion on the nH graph well within vegetation Type 1 (first 4 sample points). The compositional transition from Type 1 to Type 2 (sample points 4-9) brings on substantial rise in velocity hitting maximum at the inflexion point within Type 2, between sample points 6 and 7. From this point on, the 1st derivative levels off onto the zero velocity line.

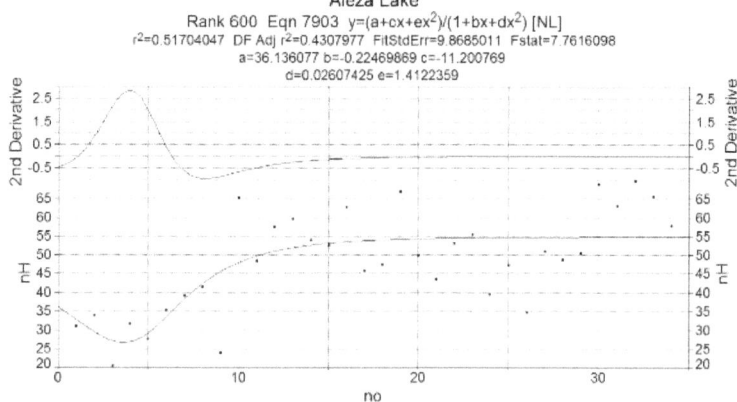

The 2nd derivative indicates maximum acceleration in the transition from Type 1 to Type 2, and maximum deceleration in the transition from Type 2 to Type 3. Thereafter the dynamic attractor's steady state sets in and nothing remarkable happens.

d. The ambitious student continues. He or she will try to rephrase the contents of paragraph 3, as appropriate, assuming that the graphs depict the local succession process. The student will argue the point in the affirmative that catenation is a valid proxy for the natural succession process and then argue the opposite.

e. The residuals are not to be ignored in any of the cases. Statistics can quantify on this basis the goodness of fit. But important to remember, that the symmetric distribution of the residuals about $f(X)$ may not indicate sole response to random causes. Therefore, the residuals should be further analysed as a data set on its own to find out what do they represent for variable X. The more ambitious student will perform the same complete analysis as before on the residuals in each case, and check what she or he can come up with.

f. I include an extra graph below: My choice of function for this is 7909 in TableCurve's list. The function gives a better fit than function 7903. It might appear to be a rather natural and better choice to capture the nH trajectory. So, the question is why do I not use Equation 7909 instead of 7903? The answer is simple. The whole thing is specious with a dip in the no = 20 to 30 interval. It is more likely that the 1957 sampling is deficient. It has left sites fitting into that interval under-sampled.

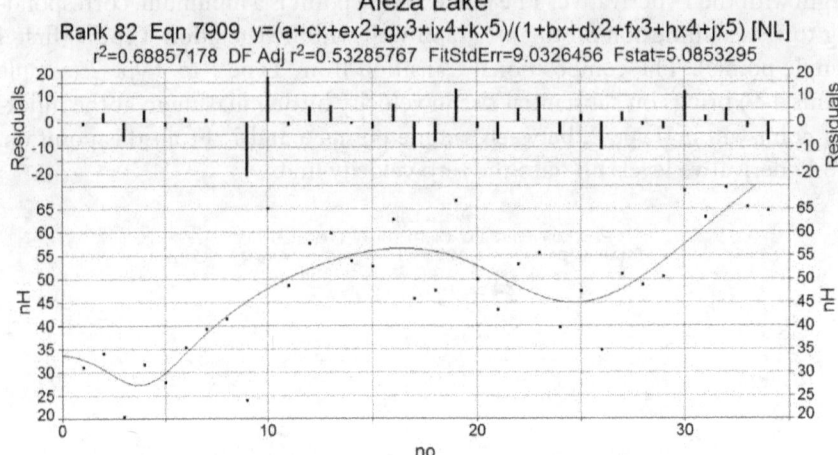

g. Turning to equation 8004 (three graphs, step 15), I remind the reader that H is the nth fraction of nH, and as such it is free from the richness effect. I let the more ambitious students offer their own evaluation. I even encourage the bravest to explore the topic on his or her field data. I am available for discussion in e-mail. The brevity of comments and questions is imperative.

5.5.3 Some questions and a step forward

I have isolated questions from correspondence about the methodology and also considered alerts regarding insufficiently presented ideas and results. I accept, the exercise is very tedious and indeed the calculus requires undivided attention to avoid mistakes. But a good number of what appears to be a discrepancy in the arithmetic can be traced back to rounding or truncation of results too soon by colleagues. Rounding errors may accumulate quickly in chain calculations, and the results do not match for different users. So, the question arose: how many valued digits to carry at each step? I say as many as the computer provides automatically. The final numbers should be rounded off at the end, and not beyond where significant digits would be lost.[16]

Comparability of EBE taken from different projects is the substance of several questions. I discussed the topic in sufficient detail in earlier publications, but I still owe explanation of a special case. This regards the comparison of nH graphs. I assume that:

a. all will use the same shape function $f(X)$;

---

[16] On the topic of significant digits, I refer the reader to the chapter on "Measurement theory" in my "Statistical Ecology" (2016).

b. all will have access to TableCurve to fit $f(X)$ and the 2nd derivative function; and

c. the computer used has capacity to keep the graphs on the monitor's screen alive, while the user accesses different applications in different windows, other than TableCurve. In what is presented in the sequel, has main concern for the characteristic points of the $f(X)$ curve, based on which results from different projects can be compared.

Processing starts with data input in TableCurve. The data and the mechanics of generating the graphs are both familiar from the preceding sections. So, we have once more the nH and 2nd derivative graphs:

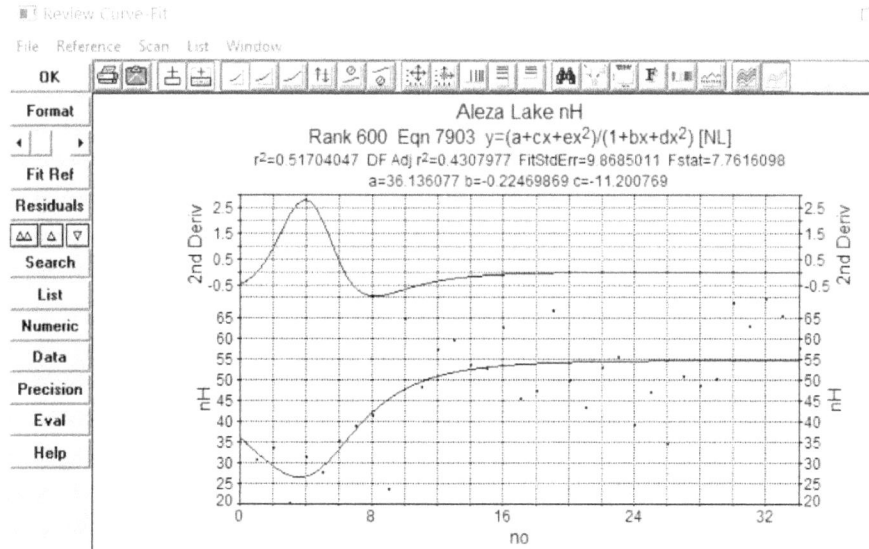

We keep this graph on the monitor's screen. Remember, "no" or alternatively X in $f(X)$ refers to sample plot position on the virtual catena (succession sere) extending from muskeg (plot 1) to the well-drained hill top (plot 34). We shall do peripheral exercises from which we return from time to time to Table-Curve's live screen graph as needed.

Our objective is to find the values which fill the cells of the following table:

2nd derivative	Global max	10% critical point	5% critical point	Locus of quasi vanishing point of the 2nd derivative	Mean nH
2nd deriv value	2.8097190743	-0.28097191	-0.140485953	-0.009982575	-0.61648907901

"no" value	3.9725260406	12.55974	14.62073	24.36	10.22043759346
nH value	26.8421108	51.6082294	52.9904599	54.6701485335	48.4733909998

We refer to this table as the "summary table". Some entries are taken from TableCurve's numeric table. Others can be read directly from the live graphs by pointing. Direct reading can give us values which are raw approximation in most cases; accuracy depends on personal skills.

To get accurate result, we have to determine critical points arithmetically by evaluation of functions or by iterations as required. We take the summary table's contents item-by-item in turn:

a. *"Global max" of the 2nd derivative.* Click "Numeric" in the graph's side bar to put the contents of the numeric table on the screen. Read the "2nd derivative max" and "X-value" in the table: 2.8097190743, 3.9725260406. Read the mean value " Ymean": 48.473390998. Symbol X stands for "no" and Y for nH.

b. *nH value at Global max.* Click on Eval in the side bar. Type 3.9725260406 into the new menu's "X box". Press "Y=Tn(X)" in the side bar. Read 26.84211083974 from the right side pane.

c. *"no" value at Average nH.* Press Eval in the side bar. Type 48.473390998 in the Y box of the new menu and press "X=Root of Y" below it. Read 10.22043759346.

d. *2nd derivative value at Mean nH.* Press Eval in the side bar. Type 10.22043759346 into the X box and click "Y=2nd Deriv at X". Read -0.61648907901.

e. *Find critical points.* These are the positional numbers, "no" values, at which the absolute value of the 2nd derivative is k% of its maximum. For example, the k10% value is 2.8097190743/10 = 0.28097190743. The k5% value is 0.28097190743/2 = 0.140485953. Find next the "no" and nH values that go with k10% and k5%. Click on Eval. Select "Generate table" in the side bar. Select "X in input, Y=2nd Derivative of Fn at X". Click O.K. Write into the boxes: Starting value 12.55, Increment 0.00001, Ending value 12.57. These are guesses based on probing the screen graph of the 2nd derivative. Read the output at line # 975:

```
973. 12.55972000000 -0.28097288702 Y=d2F(X)/dX
974. 12.55973000000 -0.28097192041 Y=d2F(X)/dX
975. 12.55974000000 -0.28097095380 Y=d2F(X)/dX
976. 12.55975000000 -0.28096998720 Y=d2F(X)/dX
977. 12.55976000000 -0.28096902060 Y=d2F(X)/dX
```

39 | Probing simple vegetation data

975 is the line where the value of the iterated 2nd derivative comes closest to the K10% value 0.28097190743 using 0.00001. We set no=12.55974. The corresponding nH value is 51.60823845344.

The values in the 5% column are similarly computed:

12072. 14.62071000000 -0.14048619303 Y=d2F(X)/dX
12073. 14.62072000000 -0.14048573523 Y=d2F(X)/dX
12074. 14.62073000000 -0.14048527743 Y=d2F(X)/dX
12075. 14.62074000000 -0.14048481963 Y=d2F(X)/dX
12076. 14.62075000000 -0.14048436183 Y=d2F(X)/dX

The value sought is in line 12074: no=14.62073. Corresponding to this is nH= 52.99045997842.

f. *Global vanishing point.* We seek the "no" value at which the absolute value of the 2nd derivative reaches minimum in the right tail of the graph. We consider the iterated results:

535. 24.34000000000 -0.01002830111 Y=d2F(X)/dX
536. 24.35000000000 -0.01000540909 Y=d2F(X)/dX
537. 24.36000000000 -0.00998257555 Y=d2F(X)/dX
538. 24.37000000000 -0.00995980034 Y=d2F(X)/dX
539. 24.38000000000 -0.00993708327 Y=d2F(X)/dX

We observe that the absolute value of the 2nd derivative is separated from its zero line by less than 0.01 value progressing right, the first time at no=24.36. I consider this as a sufficiently small value to be regarded as a quasi-vanishing point.

The final presentation should not include more than, say 4 or 5 non-zero digits. So we have the summary table:

2nd derivative	Global max	10% crit point	5% crit point	Quasi-vanishing point	Mean nH
2nd deriv's value	2.80972	-0.28097	-0.14049	-0.00998	-0.61649
no value	3.97253	12.55974	14.62073	24.36000	10.22044
nH value	26.84211	51.60823	52.99046	54.67015	48.47339

So, what do we need when we compare cases from different projects? We need to have the same shape function used, the Y variable has to be the same, and X variable must refer to a point on the same scale. Y is a universal property of things. When applied to a specific data set, the data set does not change Y, but gives it a value and a context. Y has to be interpreted accordingly. It is important to remember that the projects have to synchronise their sampling design.

5.6 Directedness

5.6.1 Preliminaries

Directedness is a property of the process trajectory (consult Orlóci 2012, 2014 and references therein). In analytical terms, the trajectory, a turning and twisting line which in sample space, connects points defined by the compositional states of stands in the same order as they are marked out by the sample plots on the catena. When we analyse succession for directedness we in fact subject succession's trajectory to scrutiny.

Considering directedness in the trajectory model context, the first thought coming to mind is the impossibility of its analytical definition in isolation from its dual, commonly known in systems theory as random walk. There are several reasons for this. At our mega level pure directedness or pure random walk does not exist. They come in convolution, but with varying levels of dominance. This is the property which we exploit in probabilistic tests, for when we test directedness, we use the zero dominance of directedness as the null model state.

We shall make directedness the object of the test in the Aleza Lake catena under the Null Hypothesis that succession is a random walk. The latter translates into the assumption that stand (community) assembly in the vegetation within which the sample plot is sited is a pure random process. This is a hypothetical condition, a unique marker for zero directedness, from which we can measure the divergence of the observed process to fix the magnitude of its directedness.

When taken in the trajectory context, directedness implies what we call a forward momentum, the opposite of back-stepping. This is easy to explain in geometric terms. Forward momentum is indicated when the inner angle of two joint segments of the trajectory, say A and B, in the A,B plain is larger than 90°. The magnitude of the momentum is the cosine of the acute angle times the length of segment B.

In any set of data there will be found a forward momentum. This is the nature of things in our ecological universe. The question is not its presence. It is there, but is its magnitude is small enough to regard it as an event commonly expected under the rule of pure chance? In other words, we have to determine the value of the forward momentum but call it significant (not trivial under) if it were an uncommonly large value under the Null Hypothesis of pure random walk. Clearly, we do not deal with absolutes, but rather with likelihoods.

After the clarifications done, we shall take a daring step not to burden further the reader with more tedious computations to complete the test. *We assume*

41 | Probing simple vegetation data

that all the cosine projections line up in sequence as we would have them in an 0-order Markov chain.

I found Markov mathematics quite powerful to interrogate the total process for directedness (of the Markov kind). I refer to the Chapter "Markov analysis" in my Statistical Ecology (Orlóci 2016 and references therein) for details regarding the methodology, and at this time I give results directly on the acceptance or rejection of the null hypothesis of having in hand a case of random walk. What do we need for the test? We need probability points to measure the observed value's commonness. Main steps:

a. Fit the Markov chain to the data set as given in the Appendix, excluding Type 5.

b) Find the tress level in the sample. We call this the observed stress level. This is proportional to the intensity of random effects.

c) Generate probability points (call them critical values) to which we compare the observed stress level to find its commonness level. The probability points are extracted from a large number (1000 in the example) of synthetic stress values generated in as many complex iteration.

Each iteration begins with a set number of interchanges of values between randomly selected pairs of cells in the data table. The interchanges destroy the original data arrangement as if stand assembly happened under conditions specified by the Null Hypothesis.

d) Perform a new Markov analysis on the manipulated data set and find the new stress value. This stress value will be specific to conditions under the Null Hypothesis. Iterate until the selected number of iterations is fulfilled.

e. Finally, determine of the commonness of the observed stress value by counting how many of the simulated stress values are less or equal to it.

f. Do data smoothing in the original table and repeat all calculations so far discussed. Smoothing implies that data duplets are replace by their average, triplets by their average, and so forth.

After each smoothing step get a new set of probability points and perform a new test of significance. Finally, you should end up with results seen in the next section.

5.6.2 Numerics

My application program FITMARKO completes the computations automatically. The processing begins with the dialogue about preferred an mandatory specifications. The specifications are stored on file:

PROGRAM: Fitmark101025
Key reference:
Orlóci, L., Anan, M. and X.S. He. 1993. Markov chain: a realistic model for temporal coenosere? Biometrie-Praximetrie 33:7-26.

Data file:
C:\Users\Laszlo\Documents\Projects in progress\Aleza Lake 1957\Markov for Aleza\ALEZA DATA 156X34.TXT
This has 156 taxa and 34 releves.
Hypothesis tested:
Ho: series is 0-order Markov (undirected); random permutation of positions used
Output files:
Stepwise transitions in file Trnsstep1000iter.tru; this has 34 sets of 156 by 156 numbers.
Global transitions in file Trnsprob1000iter.tru; this has 156 rows and 156 columns.
Markov releves in file Markdat1000iter.tru; this has at least 34 rows and 156 columns.
Transposed Markov releves in file Tmarkdat1000iter.tru; this has 156 rows and 34 columns.
Step size upper limit used: 4

Step size is optional. If left 1, the raw data set is analysed and the processing stops. If the choice is 2 or more then neighbouring cells are replaced by their averages within the rows (taxa, species) in duplets (step size 2), triplets (step size 3), and quadruplets (step size 4). A complete analysis is performed at each step size. The smoothing process begins at each step size with the raw data. The detailed results are very voluminous. A summary table is reproduced with sufficient information to test the "0-order Markov" (random walk) hypothesis:

Step size	1	2	3	4
Observed stress level	1.58288	1.45009	1.41472	1.36660
% simulated values ≤ observed value	10.5	5.1	2.6	2.8
Mean of simulated values	1.60924	1.48294	1.45251	1.40448
Standard deviation (SD, simulated values)	0.01828	0.01625	0.01507	0.01605
|Stress observed – Mean stress|/ SD *	1.44192	2.02114	2.50812	2.36011

*Standard normal variate. The density function applies.

The second row tells the story: the raw data set encapsulates weak Markov type directedness at step size one. The observed stress vale is too large (1.58288) in the sense that stress values less than 1.58288 are commonly expected under O-order Markovity. As can be seen, data smoothing reduced the observed stress sufficient to conclude that a significant directedness exists. Put it in another way, the catena captures a process in Nature which we should not

43 | Probing simple vegetation data

regard as a random walk. The following narrates the particulars up to step size 4:

PROGRAM: Fitmarko101025
Key reference:
Orlóci, L., Anan, M. and X.S. He. 1993. Markov chain: a realistic model for temporal coenosere? Biometrie-Praximetrie 33:7-26.
==
Post mortem - must be read with attention:
-- Time series data were expected. Constant time-step width was assumed. Markov transitions were computed based on population gains and losses in the releves across time steps with LAG from 1 up to 4.
Distance matrices of natural relevés and Markov relevés were compared by the Kruskal's tress index. Probabilities were determined in a Monte Carlo experiment, involving random permutation of relevé labels in Option 1 and the resampling of the stretched Markov chain in Option 2. The stretching involved Markov co-ordinates M by way of an auxiliary array u of f.. cells in such a way that the code of any specific element of M will appear in a specific number of cells in u equal to the value of that element. Symbol f.. is the grand total of all elements in M.
Regarding the Ho tested, the following are relevant:
Option 1 -- Ho: the observed series is chaotic as if it were generated by a random walk. Ho is tested against the alternative H1: the series is directed. Since Ho implies a high stress value, H1 is accepted when the stress is unusually small in probability terms.
Option 2 -- Ho: the series is m order Markov. In this case Ho is tested against the alternative hypothesis H1: the series is not m order Markov. Since Ho implies a low stress value, H1 is accepted if the stress is unusually large in probability terms.
Orlóci et al. (1993), Biom. Praxim. 33, 7-26.

Data file:
C:\Users\Laszlo\Documents\Projects in progress\Aleza Lake 1957\Markov for Aleza\
ALEZA DATA 156X34.TXT
This has 156 taxa and 34 relevés.
Hypothesis tested: Ho: series is 0-order Markov (undirected); random permutation of positions used
Output files:
Stepwise transitions in file Trnsstep1000iter.tru; this has 34 sets of 156 by 156 numbers.
Global transitions in file Trnsprob1000iter.tru; this has 156 rows and 156 columns.
Markov releves in file Markdat1000iter.tru; this has at least 34 rows and 156 columns.
Transposed Markov releves in file Tmarkdat1000iter.tru; this has 156 rows and 34 columns.
Step size upper limit used: 4. Number of iterations per step: 1000.

STEP SIZE: 1
==
Observed (original) stress value: 1.5828796
Percent of generated values less or equal to 1.5828796 is 10.5
Mean of generated stress values: 1.6092402
Variance of the generated stress values: 3.342167e-4
Standard deviation of the generated stress values: 1.8281595e-2
Variance of the generated mean: 3.342167e-7
Standard deviation of the generated mean: 5.7811478e-4
Observed stress expressed as a deviation from the generated mean in standard units: -1.4419242

Test of Ho: the coenosere is undirected (zero order Markov).
Test criterion: s(DX;DM). In this, DX is the distance configuration
of the observed relevés and DM is the distance configuration of
relevés based on the Markov scores M fitted to X. The reference
distribution is based on RNDX, the random-permuted X, and s(DRNDX;DM)
but in this case DM is the Markov distance configuration based on M
fitted to RNDX. Probabilities and probability points are given below.
The probabilities, given as % values are in the left tail of the
s(DRNDX;DM) distribution.

1%	5%	10%	25%	50%	75%	90%	95$	99%
1.55103	1.56873	1.58252	1.60128	1.61369	1.62219	1.62757	1.63089	1.63508

*The probability points are stress values. Reject the null hypothesis
when the observed stress value associates with a small probability.

STEP SIZE: 2

Observed stress value: 1.4500889
Percent of generated values less or equal to 1.4500889 is 5.1
Mean of generated stress values: 1.4829424
Variance of the generated stress values: 2.6422282e-4
Standard deviation of the generated stress values: 1.6254932e-2
Variance of the generated mean: 2.6422282e-7
Standard deviation of the generated mean: 5.1402609e-4
Observed stress expressed as a deviation from the generated mean
in standard units: -2.0211414

Test of Ho: the coenosere is undirected (zero order Markov).
Test criterion: s(DX;DM). In this, DX is the distance configuration
of the observed relevés and DM is the distance configuration of
relevés based on the Markov scores M fitted to X. The reference
distribution is based on RNDX, the random-permuted X, and s(DRNDX;DM)
but in this case DM is the Markov distance configuration based on M
fitted to RNDX. Probabilities and probability points are given below.
The probabilities, given as % values are in the left tail of the
s(DRNDX;DM) distribution.

1%	5%	10%	25%	50%	75%	90%	95$	99%
1.43330	1.44958	1.45964	1.47420	1.48604	1.49479	1.50067	1.50284	1.50715

*The probability points are stress values. Reject the null hypothesis
when the observed stress value associates with a small probability.

STEP SIZE: 3

Observed stress value: 1.4147152
Percent of generated values less or equal to 1.4147152 is 2.6
Mean of generated stress values: 1.4525102
Variance of the generated stress values: 2.270751e-4
Standard deviation of the generated stress values: 1.5069011e-2
Variance of the generated mean: 2.270751e-7
Standard deviation of the generated mean: 4.7652398e-4
Observed stress expressed as a deviation from the generated mean
in standard units: -2.5081231

Test of Ho: the coenosere is undirected (zero order Markov).
Test criterion: s(DX;DM). In this, DX is the distance configuration

of the observed relevés and DM is the distance configuration of
relevés based on the Markov scores M fitted to X. The reference
distribution is based on RNDX, the random-permuted X, and s(DRNDX;DM)
but in this case DM is the Markov distance configuration based on M
fitted to RNDX. Probabilities and probability points are given below.
The probabilities, given as % values are in the left tail of the
s(DRNDX;DM) distribution.

  1%   5%   10%   25%   50%   75%   90%   95$ 99%
1.40480 1.42270 1.43222 1.44502 1.45527 1.46382 1.46887 1.47130 1.47552
*The probability points are stress values. Reject the null hypothesis
when the observed stress value associates with a small probability.

STEP SIZE: 4

Observed stress value: 1.3665969
Percent of generated values less or equal to 1.3665969 is 2.8
Mean of generated stress values: 1.4044751
Variance of the generated stress values: 2.5757966e-4
Standard deviation of the generated stress values: 1.6049288e-2
Variance of the generated mean: 2.5757966e-7
Standard deviation of the generated mean: 5.0752306e-4
Observed stress expressed as a deviation from the generated mean
in standard units: -2.3601147

Test of Ho: the coenosere is undirected (zero order Markov).
Test criterion: s(DX;DM). In this, DX is the distance configuration
of the observed relevés and DM is the distance configuration of
relevés based on the Markov scores M fitted to X. The reference
distribution is based on RNDX, the random-permuted X, and s(DRNDX;DM)
but in this case DM is the Markov distance configuration based on M
fitted to RNDX. Probabilities and probability points are given below.
The probabilities, given as % values are in the left tail of the
s(DRNDX;DM) distribution.

  1%   5%   10%   25%   50%   75%   90%   95$ 99%
1.35691 1.37162 1.38221 1.39657 1.40728 1.41592 1.42207 1.42515 1.42962
*The probability points are stress values. Reject the null hypothesis
when the observed stress value associates with a small probability.

Just that much more that FITMARKO is my tool; I have written for my own use. Others benefited in first use from direct supervision.

6. The broader context

This essay is very much about physical structures whose causes are among the principal object of scientific research in vegetation science. I emphasise the phytosociological connection to alert the reader to my community-level approach. I add, that in the best tradition of phytosociology, the contents are very much ecological. It could not be any other way, considering that no understanding of space/time dynamics in vegetation structure and functionality is conceivable outside the conceptual framework of ecology.

Within that conceptual world, entropy is either *information-based* or *energy-based*. The model, which gives practicality to the later, is the *resonator complex model*. The model is stand (patch, community) level and plant taxon based. Information theory hands us the entropy equation $H = \sum_{i=1}^{N} p_i \ln p_i$. We borrow the energy-based entropy's equation $E = -\ln P$ from quantum theory. E is written for a set of discrete probabilities linked to n independent resonators. The information theoretical H scales disorder in the plant community. E is a proxy scaler of potential energy.

Note that in E there is a connection to the distribution of the number of energy units (discrete energy packet, quanta) among the resonators. For energy units we substitute taxon performance units. Regarding E, we do not use p_i of the informational theoretical equation of H. We use P, the probability of a vegetation stand, exactly the same as the stand actually observed, arising by pure chance.

At this point I find it important to note that the basic principles which define energy-based entropy's validity as a proxy for potential energy, have universal validity at any scale. Yet, we do not do quantum mechanics in my field for the simple reason that the techniques usually do not work on the macroscale. Consider the universal constant k in Max Plank's energy-based entropy equation[17],

$$E = k \ln W + \text{constant}$$

The value of k can be calculated at any scale, but there is no utility in this for us, because it simply vanishes into insignificance in macro scale applications. Regarding the forgoing, the fact remains that entropy, energy, and probability are linked, and energy-based entropy is proportional to the unmeasurable potential energy.

It is customary to use models when we deal with specific properties of processes. Models are like crutches, so to speak, that help as walk without limping through the problem for a price. And this may be substantial. The trick is to find the model which can serve the purpose without translating the problem into something that were not intending to research. The Markov chain is a case in point. Was it good choice? It was a reasonable one for the question posed. We asked a simple question: Does the succession process have a directedness of the Markov type? When we concluded with a "YES", our case became a member of the general class known as Markovian

---

[17] In our notations $W = \dfrac{1}{P}$

processes. Why is this an advantage? Simply, all information about the class applies to the individual cases. Based on which we generalise our findings and establish connection.

In closing, I have to dispel two misconceptions:

Misconception 1. *Disorder and its measurement is inseparably linked to count or frequency data.* This goes against common sense. I suggest that the choice of the disorder generating trait is a prerogative for the user, but this must be explicitly stated in the interpretation of the results.

Misconception 2. *When the expression* E = - ln P *is identified as a proxy measure for potential energy, the term 'energy' is a metaphor.* Knowing that we cannot measure energy directly, and we cannot use the constant k, what remains to be measured are the manifestations of energy. Then logic dictates that in all cases or in none of the cases is the term 'energy' a metaphor.

Misconception 3. *Application of the Markov chain is dependent on a priory theory based transition probabilities.* It is unlikely that dealing with a complex natural process, such as plant community succession, we can find much help from that direction. But we can go a different way and rely on empiricism. I suggested a practical and effective method to determine transition probabilities from survey type vegetation data. I already referred to sources. We used the relevant algorithm in the example already discussed.

References

Arlidge J.W.C. 1952. Ecological investigations in the Spruce-Alpine fir type, Aleaza Lake Experimental Forest, British Columbia Forest Service, Research Division, Victoria, B.C, Canada.

Arlidge J.W.C. 1956. Ecological investigations in the Spruce-Alpine fir type, Aleza Lake Experimental Forest, British Columbia Forest Service, Research Division, Victoria, B.C, Canada.

Clements, F.E. 1916. Plant Succession: an Analysis of the Development of Vegetation. Publ. No. 242, Carnegie Institution, Washington.

Clements, F.E. 1936. Nature and structure of the climax. J. Ecol. 24: 252-284.

Cooper, W.S. 1926. The fundamentals of vegetation change. Ecology 7:391-413.

Cowles, H.C. 1899. The ecological relations of the vegetation on the sand dunes of Lake Michigan. Part. I. Geographical relations of the dune floras. Bot. Gaz. 27: 95-117.

Decie, T.P. 1956. Working plan the Forest Experimental Station, Aleza Lake. British Columbia Forest Service, Research Division, Victoria, B.C, Canada.

Fraser, A. R. and J.L. Alexander. 1949. Development of the Spruce-Balsam type in the Aleza Lake Experimental Forest. British Columbia Forest Service, Research Division, Victoria, B.C, Canada.

Gleick, J. 1987. Chaos. Making a new science. Pinguin Books, England.

Griffith, B.G. 1926. Herbarium collection at the Forest Experimental Station at Aleza Lake. British Columbia Forest Service, Research Division, Victoria, B.C., Canada.

Halliday, W.E.D. 1937. Forest classification of Canada. Forest Service Bulletin 89, Ottawa. http://publish.uwo.ca/~lorloci/Koa/Syndynamics.pdf

Hult, R. 1881. Fors $\bar{k}$ til analytiskbehandling af växformationerna. Medd. Soc. Faun. Flor. Fenn. 8. -- 1885. Blekinges vegetation. Ett bidrag till växformationernas. Medd. Soc. Faun. Flor. Fenn. 8.

Kelley, C.C. and L. Farstad. 1946. Soil survey of the Prince George area, British Columbia. British Columbia Forest Service, Research Division, Victoria, B.C., Canada.

Kerner von Marilaun, A. 1863. Das Pflanzenleben der Danauländer. Innbruck, Wagner. - Conard, H.S. 1951. The Background of Plant Ecology. The Iowa State University Press, Ames (1977, Arno Press, New York.) – This is a translation of Kerner von Merilaun's "Das Pflanzenleben der Donauländer" under a mor fitting Engkish title.

Krajina, V.J. 1965. Biogeoclimatic zones and classification of British Columbia. In: Ecology of Western North America, Vol. 1: 1-17. Department of Botany, British Columbia, Vancouver, Canada.

Kujala, V. 1945. Waldvegetation Untersuchungen in Canada. Anales Academicae Scientiarum Fennica. Ser. A. IV, Biologia, Helsinki.

Orlóci, L. 1958. Data for the classification of plant communities at Eliza Lake, Central British Columbia. Department of Botany, University of British Columbia, Vancouver, B.C, Canada.

Orlóci, L. 1965. The coastal western hemlock zone on the South-Western British Columbia Mainland. In: Ecology of Western North America, Vol. 1: 18-34. Department of Botany, British Columbia, Vancouver, Canada.

Orlóci, L. 1993. The complexities and scenarios of ecosystem analysis. *In*: G. P. Patil and C. R. Rao, Multivariate Environmental Statistics, pp.421-430, North Holland/Elsevier, New York.

Orlóci, L. 2000. From Order to Causes. A personal view, concerning the principles of syndynamics. Download file 'Syndynamics', URL.

Orlóci, L. 2001. Prospects and expectations: reflections on a science in change. Community Ecology 2: 187-196.

Orlóci, L. 2006. Diversity partitions in 3-way sorting: functions, Venn diagram mappings, typical additive series, and examples. Community Ecology 7:253-259.

Orlóci, L. 2012.Statistical multiscaling in dynamic ecology. Probing the long-term vegetation process for patterns of parameter oscillations. SCADA Electronic Books. CreateSpace eStore: https://www.createspace.com/3830594 .

Orlóci, L. 2014. Statistical Ecology. The quantitative exploration of nature to reveal the unexpected. SCADA Publishing, Canada. Online Edition: https://createspace.com/3476529 .

Orlóci, L. 2015. Diversity analysis, holistic energetics, and statistics: The resonator complex model of the vegetation stand. SCADA Publishing, Canada. Online edition: https://www.createspace.com/5783923

Orlóci, L. 2016. Statistical quantum ecology. Essays on the resonator complex model of the vegetation stand . SCADA Publishing, Canada. Online Edition: https://createspace.com/ 6509504

Peck, M.E. 1961. A manual of the higher plants of Oregon. Binford & Mort Publishing, Hillsboro, Oregon.

Spilsbury, R. H. and Smith, D. S. 1947. Forest site types of the Pacific Northwest. British Columbia Forest Service, Research Division, Victoria, B.C., Canada.

Taylor, T.M.C. 1963. The Ferns and Fern-allies of British Columbia. British Columbia Provincial Museum. Ottawa, Canada.

Warming, E. 1895. Plantesamfund. Grundtraek of den økologiske Plantegeografi. Kjøbenhavn, Philipsens.

Index

"no" value, 38, 39
1957 essay, 6
2nd derivative, 30, 34, 35, 37, 38, 39
Alder, 11, 53
Alexander, 10
alluvia, 9, 10
Alnus, 11, 53, 56, 58
amount of energy, 7

application program, 42, 43
Aralia, 11, 53, 56, 58
Arlidge, 6, 10, 47
Arlidge J.W.C., 47
Arlidge,, 6, 10
Art Prochnau, 6
assembly/disassembly, 7, 8, 21
Athyrium, 10, 11, 53, 56, 58
attractor, 8, 35

back-stepping, 40
Beringia, 4
binary effect, 14
Black spruce, 11, 53
Boron River, 1, 10, 11
British Columbia, 1, 6, 8, 10, 47, 48, 49
CaptureWizPro, 25, 30
Caribou Park Land, 10
catena, 18, 20, 21, 22, 34
Catena, 18, 22
chance, 15
characteristic points, 37
chronosere, 4, 8
Cladonia, 12, 54, 56, 59
Clements, F.E., 47
climatope, 8
climax, 8, 11, 17, 47
Columbia Cedar-Hemlock Forest, 10
commonness, 15, 41
community development, 8, 21, 22
Comparability, 36
complex, 15
composition, 8, 16, 20, 22
Cooper, W.S., 47
Cornus, 11, 54, 56, 59
correlation, 14
Corylus, 12, 54, 56, 59
covariance, 14
cover, 7, 12, 13, 14, 15, 16, 53
cover values, 13
Cowles, H.C., 47
critical points, 38
crown canopy, 13, 53
Darwin-Wallis theory, 21
data arrangement, 41
data input, 37
data set, 6, 7, 8, 14, 35, 53

Decie, T.P., 47
density function, 42
Derivative-calculator, 25, 31
differential equation, 25
directedness, 40, 41, 42
disorder generating trait, 47
Disporum, 11, 54, 56, 59
disturbance regime, 8
Dokuchaev's theory, 21
Dryopteris, 11, 54, 56, 59
dynamic structures, 20
EBE, 15
EBE instability, 15
ecology, 4
edaphotope, 8
Eigenanalysis, 14
eigenvalues, 14
eigenvectors, 14
emergent energy-based entropy, 17
energy, 7, 15, 16
energy level, 7, 15, 16, 17
energy structure, 4, 16
energy unit, 15, 16
energy-based entropy, 7, 14, 16, 46
Entodon, 11, 12, 54, 56, 59
entropy, 46
environmental effects, 8
environmental mediation, 7, 17, 19
Epilobium, 12, 54, 56, 59
error condition, 24
error generating events, 7
errors, 24, 36
Example, 17, 25
facilitation, 8, 21, 22
feedback, 8
Ferenc Tuskó, 2, 6
Fisher, 22

FITMARKO, 42
flipping, 15
floodplain forest, 4
forward momentum, 40
Fraser, 10, 48
Fraser, A. R., 48
ghost energy-based entropy, 17
ghost states, 15, 16
glacial outwash, 9
Gleick, 21
Gleick, J., 48
Global max, 37, 38, 39
Griffith, 10, 48
Griffith, B.G., 48
ground cover, 12, 13
György Leskó, 6
Halliday, W.E.D., 48
herb layer, 13, 14, 53
homogeneity, 8, 17
Homogeneity, 8
Hult, R., 48
Hylocomnium, 11, 12
information, 4
information potential, 1, 5
information richness, 6
information-based entropy, 46
instability, 15, 16, 17
iterated results, 39
iteration, 19, 20, 41
J.L. Alexander, 48
Kalmia, 11
Kelley, C.C., 48
Kerner, 8, 20, 21, 22, 48
Kerner von Marilaun, 8, 20, 48
Kerner von Marilaun, A., 48
Kernerian, 21, 22
Krajina, V.J., 48
Kujala, 10, 11, 48
Kujala, V., 48
Ledum, 11, 54, 57, 59

logistic functions, 21
Lotka, 21
Lotka-Volterra, 21
Markov analysis, 41
Markov chain, 41, 42, 43
Markov mathematics, 41
Matteucia, 11
Max Planck, 14, 15, 16
mean value, 38
Mendel's theory, 21
mosaic, 8
multiscaling, 4
nats, 15, 16, 18
nH series, 24
nH value, 38, 39
Null Hypothesis, 40, 41
numeric table, 38
Numerics, 25
Oplopanax, 11, 54, 57, 59
organic deposits, 9
Orlóci, 3, 4, 6, 14, 19, 22, 40, 41, 42, 43, 48, 49, 53
Orlóci, L., 4, 48, 49
patches, 8
peatbog, 20
Peck, M.E., 49
pedogenesis, 21
Peltigera, 12, 55, 57, 59
Peter Greig-Smith, 6
phylogenetic process, 7
phytosociological data, 1, 5, 6, 14
phytosociological gestaltism, 5
phytosociological synthesis, 6, 8
Picea, 11, 12, 55, 57, 60
Ponderosa pine savannah, 10
Populus, 12, 55, 57, 60
potential energy structure, 14, 16
Prince George, 8, 10, 48

probability, 15, 16, 18, 22, 46
probability points, 41, 44, 45
process, 4, 7, 8, 17, 20, 21, 22, 35
Proxy scalars, 15
proxy succession, 22
pure directedness, 40
pure random walk, 40
Quantum analysis, 4
Quantum Ecology, 4
random, 8, 15, 16, 17, 19, 35
random process, 40
random walk, 40, 41, 42
resonator complex, 4, 14, 15, 16, 28, 46, 49
Rhytidiadelphus, 11, 55, 57, 60
richness, 7, 12, 13, 14, 19, 34, 36
Road map, 10
sample plot, 7, 8, 13, 34
shape function, 24, 25, 36, 39
shrub, 10, 13, 14, 53
Sopron Division, 6
Sphagna, 11
Sphagnum, 11, 53, 55, 58, 60
Sphagnum moss, 11, 53
Spilsbury, R., 49
spreadsheet, 25
Spruce Forest, 10
stability, 15, 16
stand, 15, 16
stand elements, 8
standard deviation, 16
Standard normal variate, 42
stand-level structure, 7
Statistical Ecology, 4, 36, 41, 49
statistical practice, 7
statistical significance, 15
stress, 41, 42, 43, 44, 45
structural trait, 6, 7
structure, 4, 7, 8, 13, 14, 16, 17, 23, 47
Struts fern, 11, 53
Subalpine fir, 11, 12, 53
Sub-Boreal, 6, 10
succession, 4, 7, 8, 11, 20, 21, 22, 35
succession sere, 22, 37
succession's trajectory, 40
summary table, 38, 42
TableCurve, 24, 25, 26, 30, 31, 32, 33, 34, 35, 37, 38
taxon cover, 14
taxon richness, 14
Taylor, T.M.C. 1963., 49
test of significance, 41
Thomas M.C. Taylor, 6
time series, 8, 34
traits, 1, 4, 5, 6, 7, 14
trajectory, 24, 35, 40
Type 1, 10, 11, 12, 17, 18, 35, 53
Type 2, 11, 18, 35, 53
Type 4, 12, 53, 58
unit instability moment, 16
Urtica, 11, 55, 58, 60
V.J. Krajina, 6
Vaccinium, 10, 11, 12, 55, 58, 60
vanishing point, 37, 39
vegetation, 1, 4, 5, 6, 7, 8, 10, 11, 12, 13, 14, 15, 16, 17, 20, 21, 22, 28, 34, 35, 45, 46, 47, 48, 49
vegetation process, 4
vegetation stand, 4, 7, 16, 28, 46, 49
vegetation structure, 6, 16, 46
vegetation type, 7, 8, 12, 13, 15, 17, 20
vegetation types, 6, 11, 14

virtual catena, 37
virtual sere, 18
visible object, 13
Vladimir J. Krajina, 6, 9
Volterra, 21

Warming, E., 49
White spruce, 11, 12, 53
Willow, 12, 53
zero directedness, 40

Appendix

The data set includes cover estimates for 156 taxa in 37 sample plots from the original source (Orlóci 1957). These describe the five forest types borrowed from the 1957 report:

Type 1. Black spruce – Sphagnum moss
Type 2. Alder –Struts fern
Type 3. White spruce - Subalpine fir
Type 4. Lodge pole pine - Lichen
Type 5. Willow - Willowherb

The entries in the tables are cover estimates on the 5-state phytosociological scale: 1 1-5%, 2 5-25%, 3 25-50%, 4 50-75%, 5 75-100%. Letter L stands for layer in the sampled stand: A for crown canopy, B for the shrub layer, C for the herb layer, and D for the ground of mosses, liverworts, and lichens.

Blocks in the table represent the types. Overall order of relevés follow the estimated soil moisture gradient, except Type 5.

Aleza Lake data tables:

Taxa	#	L	Type 1				Type 2				
			1	2	3	4	1	2	3	4	5
Abies lasiocarpa	1	A	1	0	0	0	0	0	0	0	0
Abies lasiocarpa	2	C	0	0	0	0	0	0	0	0	0
Abies lasiocarpa	3	B	0	1	0	0	0	0	0	0	0
Acer glabrum	4	B	0	0	0	0	1	0	0	0	0
Aconitum columbianum	5	C	0	0	0	0	0	0	0	0	0
Actaea spicata	6	C	0	0	0	0	0	0	0	0	0
Agropyron sp.	7	C	0	0	0	0	0	0	0	0	0
Alnus synuata	8	B	0	0	0	0	0	0	0	0	0
Alnus tenuiflora	9	B	0	0	0	0	0	0	0	0	0
Alnus tenuifolia	10	B	0	0	1	0	4	5	3	3	5
Amelanchier alnifolia	11	B	0	0	0	0	0	0	0	0	0
Anaphalis margaretacea	12	C	0	0	0	0	0	0	0	0	0
Andromeda polyfolia	13	B	1	0	0	1	0	0	0	0	0
Aralia nudicaulis	14	C	0	0	0	0	0	0	0	0	0
Arnica cordifolia	15	C	0	0	0	0	0	0	0	0	0
Aruncus silvester	16	C	0	0	0	0	0	0	1	1	0
Asarum caudatum`	17	C	0	0	0	0	0	0	0	0	0
Aster douglasii	18	C	0	0	0	0	0	0	0	0	0
Aster sp.	19	C	0	0	0	0	1	1	0	0	0
Athyrium filix-femina	20	C	0	0	0	0	0	11	2	1	0
Betula glandulosa	21	B	0	0	1	3	0	0	0	0	0
Betula papyrifera	22	A	0	0	0	0	0	0	0	0	0

Betula papyrifera	23	B	0	0	0	0	0	0	0	0	0
Brachypodium sp.	24	C	0	0	0	0	0	0	0	0	0
Calamagrostis sp.	25	C	0	0	0	0	0	0	0	0	0
Carex lenticularis	26	C	1	1	1	2	0	0	0	0	0
Carex microglochin	27	C	1	0	0	1	0	0	0	0	0
Carex sp.	28	C	0	0	0	0	0	0	0	0	0
Carex tetanica	29	C	1	1	1	1	0	0	0	0	0
Carex trisperma	30	C	0	1	1	0	0	0	0	1	0
Chimaphylla umbellata	31	C	0	0	0	0	0	0	0	0	0
Chiogenes hispidula	32	C	1	1	0	2	0	0	0	0	0
Cinna latifolia	33	C	0	0	0	0	1	0	0	0	1
Circaea alpina	34	C	0	0	0	0	1	1	0	0	0
Cladonia rangiferina	35	D	0	0	0	0	0	0	0	0	0
Cladonia sp.	36	D	0	1	0	0	0	0	0	0	0
Cladonia sylvatica	37	D	0	0	0	0	0	0	0	0	0
Clintonia uniflora	38	C	0	0	0	0	0	0	0	0	0
Comandra umbellata	39	C	0	0	0	0	0	0	0	0	0
Comarum palustre	40	C	0	1	0	1	0	0	0	0	0
Cornus canadensis	41	C	0	0	0	0	0	0	0	0	0
Cornus stolonifera	42	B	0	0	0	0	0	0	0	0	1
Corylus cornuta	43	B	0	0	0	0	0	0	0	0	0
Dicranum scoparius	44	D	0	0	0	0	0	0	0	0	0
Disporum oreganum	45	C	0	0	0	0	0	1	0	0	0
Dryopteris phegopteris	46	C	0	0	0	0	0	0	0	0	0
Dryopteris austriaca	47	C	0	0	0	0	0	0	1	1	0
Dryopteris sp.	48	C	0	0	0	0	0	0	0	0	0
Dryopteris disjuncta	49	C	0	0	0	0	0	0	1	1	0
Dryopteris filix-mas	50	C	0	0	0	0	0	0	0	0	0
Elymus sp.	51	C	0	0	0	0	0	0	0	0	0
Entodon schreberii	52	D	0	1	0	1	0	0	0	0	0
Epilobium adenocaulon	53	C	0	0	0	0	0	1	0	0	0
Epilobium angustifolium	54	C	0	0	0	0	0	0	1	0	0
Equisetum fluviatilis	55	C	1	0	0	1	0	0	0	0	0
Equisetum pratense	56	C	1	1	0	0	1	2	2	1	1
Equisetum sylvaticum	57	C	2	2	0	0	1	1	2	1	0
Fragaria sp.	58	C	0	0	0	0	0	0	0	0	0
Fritillaria lanceolata	59	C	0	0	0	0	0	0	1	1	0
Galeopsis tetrachit	60	C	0	0	0	0	0	0	0	0	0
Galium aparine	61	C	0	0	0	0	1	1	1	1	1
Galium borealis	62	C	0	0	0	0	0	0	0	0	0
Galium triflorum	63	C	0	0	0	0	0	0	1	0	0
Geranium sp.	64	C	0	0	0	0	0	0	0	0	0
Geum oreganum	65	C	0	0	0	0	1	1	1	0	0
Habenaria sp.	66	C	0	0	0	0	0	0	0	0	0
Heracleum lanatum	67	C	0	0	0	0	1	1	2	1	1
Hylocomium splendens	68	D	0	0	0	0	0	0	0	0	0
Hypnum sp.	69	D	0	0	0	0	0	0	0	0	0
Impatiens biflora	70	C	0	0	0	0	1	1	1	4	0
Juniperus communis	71	B	0	0	0	0	0	0	0	0	0
Ledum groenlandicum	72	B	1	2	1	1	0	0	0	0	0
Larix laricina	73	A	0	0	0	1	0	0	0	0	0
Larix laricina	74	B	0	0	0	1	0	0	0	0	0
Lathyrus ochroleucus	75	C	0	0	0	0	0	0	0	0	0
Linnaea borealis	76	C	0	0	0	0	0	0	0	0	0
Listera cordate	77	C	0	1	0	0	0	0	0	0	0
Lonicera involucrata	78	B	0	0	0	0	1	1	2	4	1
Lycopodium annotinum	79	C	0	0	0	0	0	0	0	0	0
Lycopodium clavatum	80	C	0	0	0	0	0	0	0	0	0
Lycopodium complanatum	81	C	0	0	0	0	0	0	0	0	0
Lycopodium obscurum	82	C	0	0	0	0	0	0	0	0	0
Lyschitum americanum	83	C	1	1	0	0	0	0	0	0	0
Maianthemum uniflorum	84	C	0	0	0	0	0	0	0	0	0
Matteuccia struthiopteris	85	C	0	0	0	0	5	3	3	2	5
Melampyrum sp.	86	C	0	0	0	0	0	0	0	0	0
Mitella nuda	87	C	0	0	0	0	0	0	0	0	0

Probing simple vegetation data

Species	#										
Mnium insigne	88	D	0	0	0	0	0	0	0	0	0
Mnium punctatum	89	D	0	0	0	1	0	0	0	0	0
Moneses uniflora	90	C	0	0	0	0	0	0	0	0	0
Oplopanax horridus	91	B	0	0	0	0	0	0	0	0	0
Orysopsis asperifolia	92	C	0	0	0	0	0	0	0	0	0
Peltigera sp.	93	D	0	0	0	0	0	0	0	0	0
Petasites speciosa	94	C	0	0	0	0	0	0	0	0	0
Picea engelmanii	95	A	0	0	0	0	0	0	0	0	0
Picea glauca	96	A	0	0	0	0	0	0	0	0	0
Picea glauca	97	B	0	0	0	0	0	0	0	0	0
Picea glauca	98	C	0	0	0	0	0	0	0	0	0
Picea mariana	99	A	3	3	3	5	0	0	0	0	0
Picea mariana	100	B	2	1	1	1	0	0	0	0	0
Picea mariana	101	C	0	0	0	0	0	0	0	0	0
Pinus contorta v. latifolia	102	A	1	0	0	0	0	0	0	0	0
Pinus contorta v. latifolia	103	B	0	0	0	0	0	0	0	0	0
Plagiothecium undulatum	104	D	0	0	0	0	0	0	0	0	0
Poa sp.	105	C	1	0	0	0	0	0	0	0	1
Polytrichum juniperinum	106	D	1	0	1	0	0	0	0	0	0
Populus X	107	A	0	0	0	0	0	0	0	0	0
Populus tremuloides	108	B	0	0	0	0	0	0	0	0	0
Populus tremuloides	109	A	0	0	0	0	0	0	0	0	0
Populus trichocarpa	110	A	0	0	0	0	0	0	0	1	0
Pseudotsuga menziesii	111	B	0	0	0	0	0	0	0	0	0
Pseudotsuga menziesii	112	A	0	0	0	0	0	0	0	0	0
Pteridium aquilinum	113	C	0	0	0	0	0	0	0	1	0
Pyrola asarifolia	114	C	0	0	0	0	0	0	0	0	0
Pyrola secunda	115	C	0	1	0	0	0	0	0	0	0
Ranunculus repens	116	C	0	0	0	0	0	0	1	0	1
Rhodobryum roseum	117	C	0	0	0	0	0	0	0	0	0
Rhytidiadelphus triquetus	118	D	0	0	0	0	0	0	0	0	0
Ribes hudsoniana	119	B	0	0	0	0	0	0	0	0	1
Ribes lacustre	120	B	0	0	0	0	0	0	0	1	0
Rosa sp.	121	B	0	0	1	0	0	0	0	0	0
Rubus ideus	122	B	0	0	0	0	0	1	0	0	0
Rubus parviflorus	123	B	0	0	0	0	0	0	0	1	0
Rubus pedatus	124	C	0	0	0	0	0	0	0	0	0
Rubus sp.	125	B	0	0	0	0	0	0	0	0	0
Salix sp.	126	B	0	1	1	0	0	0	0	1	0
Sambucus pubens	127	B	0	0	0	0	1	1	1	1	0
Sambucus sp.	128	B	0	0	0	0	0	0	0	0	0
Smilacina racemosa	129	C	0	0	0	0	0	0	1	1	0
Sonchus asper	130	C	0	0	0	0	1	0	0	1	1
Sorbus sitchensis	131	B	0	0	0	0	0	0	0	0	0
Sphagnum sp.	132	D	5	5	3	5	0	0	0	0	0
Spiraea densiflora	133	B	0	0	0	0	0	0	0	0	0
Spiraea lucida	134	B	0	0	0	0	0	0	0	0	0
Spiraea menziesii	135	B	1	0	0	0	0	0	0	0	0
Spiraea pyramidata	136	B	0	0	0	0	0	0	0	0	0
Sreptopus amplexifolius	137	C	0	0	0	0	0	1	1	0	1
Streptopus sp.	138	C	0	0	0	0	0	0	0	0	0
Streptopus roseus	139	C	0	0	0	0	1	0	1	1	0
Symphoricarpus albus	140	B	0	0	0	0	0	0	0	0	0
Thalictrum occidentale	141	C	0	0	0	0	0	0	0	1	0
Tiarella unifoliata	142	C	0	0	0	0	0	0	1	1	0
Trientalis latifolia	143	C	1	0	0	0	0	0	0	0	0
Tsuga heterophylla	144	A	0	0	0	0	0	0	0	0	0
Tsuga heterophylla	145	B	0	1	0	0	0	0	0	0	0
Urtica Lyalii	146	C	0	0	0	0	1	1	1	1	0
Vaccinium caespitosum	147	B	0	0	0	0	0	0	0	0	0
Vaccinium membraneaceum	148	B	0	1	0	0	0	0	0	0	0
Vaccinium occidentale	149	B	0	0	0	0	0	0	0	0	0
Vaccinium ovalifolium	150	B	0	1	0	0	0	0	0	0	0
Vaccinium oxicoccus	151	B	0	0	1	1	0	0	0	0	0
Vaccinium scoparium	152	B	0	0	0	0	0	0	0	0	0

Taxa	#	L									
Vaccinium vitis-idea	153	B	0	0	0	1	0	0	0	0	0
Veratrum viride	154	C	0	0	0	0	0	0	0	0	0
Viburnum parviflorum	155	B	0	0	0	0	0	0	0	0	0
Viola glabella	156	C	0	0	0	0	0	1	2	0	1

Type 3

Taxa	#	L	11	12	13	14	15	16	17	18	19	20
Abies lasiocarpa	1	A	2	3	4	3	3	3	4	3	3	4
Abies lasiocarpa	2	C	0	0	1	0	0	1	1	1	1	0
Abies lasiocarpa	3	B	1	1	0	1	1	1	2	0	1	1
Acer glabrum	4	B	0	0	0	0	0	0	0	0	1	0
Aconitum columbianum	5	C	0	0	0	0	0	0	0	1	0	0
Actaea spicata	6	C	1	1	1	0	0	0	0	1	0	0
Agropyron sp.	7	C	0	0	0	0	0	0	0	0	0	0
Alnus synuata	8	B	0	0	0	0	0	0	2	0	0	0
Alnus tenuiflora	9	B	0	0	0	1	0	1	0	1	0	0
Alnus tenuifolia	10	B	0	0	0	0	0	0	0	0	0	0
Amelanchier alnifolia	11	B	0	0	0	0	0	0	0	0	1	1
Anaphalis margaretacea	12	C	0	0	0	0	0	0	0	0	0	0
Andromeda polyfolia	13	B	0	0	0	0	0	0	0	0	0	0
Aralia nudicaulis	14	C	1	1	1	0	1	0	0	0	3	2
Arnica cordifolia	15	C	0	0	0	0	0	0	0	0	0	0
Aruncus silvester	16	C	0	0	0	0	0	0	0	0	0	0
Asarum caudatum`	17	C	0	0	0	0	0	0	0	0	0	0
Aster douglasii	18	C	0	0	0	0	0	0	0	0	1	0
Aster sp.	19	C	0	0	0	0	0	0	0	0	0	0
Athyrium filix-femina	20	C	1	0	1	1	1	0	0	0	0	0
Betula glandulosa	21	B	0	0	0	0	0	0	0	0	0	0
Betula papyrifera	22	A	1	0	1	1	1	1	0	0	0	0
Betula papyrifera	23	B	0	0	0	0	0	0	0	0	0	0
Brachypodium sp.	24	C	0	0	0	0	0	0	0	0	0	0
Calamagrostis sp.	25	C	0	0	0	0	0	1	0	0	0	0
Carex lenticularis	26	C	0	0	0	0	0	0	0	0	0	0
Carex microglochin	27	C	0	0	0	0	0	0	0	0	0	0
Carex sp.	28	C	0	0	0	0	0	1	0	0	0	0
Carex tetanica	29	C	0	0	0	0	0	0	0	0	0	0
Carex trisperma	30	C	0	0	0	0	0	0	0	0	0	0
Chimaphylla umbellata	31	C	0	0	0	0	0	0	0	0	1	0
Chiogenes hispidula	32	C	0	0	0	0	0	1	0	0	0	0
Cinna latifolia	33	C	1	1	1	1	0	0	0	1	0	1
Circaea alpina	34	C	1	1	1	1	0	0	0	0	0	0
Cladonia rangiferina	35	D	0	0	0	0	0	0	0	0	0	0
Cladonia sp.	36	D	0	0	0	0	0	0	0	0	0	0
Cladonia sylvatica	37	D	0	0	0	0	0	0	0	0	0	0
Clintonia uniflora	38	C	0	0	1	0	1	0	0	1	2	1
Comandra umbellata	39	C	0	0	0	0	0	0	0	0	0	0
Comarum palustre	40	C	0	0	0	0	0	0	0	0	0	0
Cornus canadensis	41	C	1	1	1	1	1	1	1	2	0	1
Cornus stolonifera	42	B	0	1	1	0	0	0	0	1	1	1
Corylus cornuta	43	B	0	0	0	0	0	0	0	0	0	0
Dicranum scoparius	44	D	0	0	1	0	0	0	1	0	1	0
Disporum oreganum	45	C	0	0	0	0	1	0	0	0	1	0
Dryopteris phegopteris	46	C	0	0	0	1	0	1	0	0	0	0
Dryopteris austriaca	47	C	1	2	2	2	1	1	0	1	0	1
Dryopteris sp.	48	C	0	0	0	0	0	0	0	0	0	0
Dryopteris disjuncta	49	C	1	2	1	2	2	1	1	2	1	1
Dryopteris filix-mas	50	C	2	0	0	0	1	1	0	1	0	0
Elymus sp.	51	C	0	0	0	0	0	0	0	0	0	0
Entodon schreberii	52	D	0	0	0	0	0	0	0	0	0	0
Epilobium adenocaulon	53	C	0	0	0	0	0	0	0	0	0	0
Epilobium angustifolium	54	C	0	0	0	0	0	0	0	0	0	0
Equisetum fluviatilis	55	C	0	0	0	0	0	0	0	0	0	0
Equisetum pratense	56	C	1	1	0	2	0	0	0	1	0	1

57 | Probing simple vegetation data

Species	#	Code										
Equisetum sylvaticum	57	C	1	2	1	1	0	3	0	1	0	1
Fragaria sp.	58	C	0	0	0	0	1	0	0	0	0	0
Fritillaria lanceolata	59	C	0	0	0	0	0	0	0	0	0	0
Galeopsis tetrachit	60	C	0	0	0	0	0	0	0	0	0	0
Galium aparine	61	C	0	0	0	0	0	0	0	0	0	0
Galium borealis	62	C	0	0	0	0	0	0	0	0	0	0
Galium triflorum	63	C	0	0	0	0	0	0	0	0	0	0
Geranium sp.	64	C	0	0	0	0	0	0	0	0	1	0
Geum oreganum	65	C	0	0	0	0	0	0	0	0	0	0
Habenaria sp.	66	C	0	0	0	0	0	0	0	0	0	0
Heracleum lanatum	67	C	0	0	0	0	0	0	0	0	0	0
Hylocomium splendens	68	D	0	0	0	1	0	2	1	1	1	0
Hypnum sp.	69	D	1	1	1	1	1	1	2	0	1	1
Impatiens biflora	70	C	0	0	0	0	0	0	0	0	0	0
Juniperus communis	71	B	0	0	0	0	0	0	0	0	0	0
Ledum groenlandicum	72	B	0	0	0	0	0	0	0	0	0	0
Larix laricina	73	A	0	0	0	0	0	0	0	0	0	0
Larix laricina	74	B	0	0	0	0	0	0	0	0	0	0
Lathyrus ochroleucus	75	C	0	0	0	0	0	0	0	0	0	0
Linnaea borealis	76	C	0	0	0	0	0	0	1	1	1	0
Listera cordate	77	C	0	0	0	0	0	0	0	0	0	0
Lonicera involucrata	78	B	1	1	0	1	1	1	1	0	0	1
Lycopodium annotinum	79	C	0	1	1	0	0	0	1	2	1	0
Lycopodium clavatum	80	C	0	0	0	0	1	0	0	0	0	0
Lycopodium complanatum	81	C	1	0	0	0	0	0	0	0	0	0
Lycopodium obscurum	82	C	0	0	0	0	0	0	0	0	0	0
Lyschitum americanum	83	C	0	0	0	0	0	0	0	0	0	0
Maianthemum uniflorum	84	C	0	0	0	0	0	0	0	0	0	0
Matteuccia struthiopteris	85	C	0	0	0	0	0	0	0	0	0	0
Melampyrum sp.	86	C	0	0	0	0	0	0	0	0	0	0
Mitella nuda	87	C	1	1	1	1	1	0	1	1	1	1
Mnium insigne	88	D	0	0	1	0	0	0	0	0	0	0
Mnium punctatum	89	D	1	1	0	1	0	0	1	1	0	0
Moneses uniflora	90	C	0	0	0	0	0	0	1	0	0	0
Oplopanax horridus	91	B	5	5	5	2	5	1	0	4	0	4
Orysopsis asperifolia	92	C	0	0	0	0	0	0	0	0	0	0
Peltigera sp.	93	D	0	0	0	0	0	0	0	0	0	0
Petasites speciosa	94	C	0	0	0	0	0	0	1	0	0	0
Picea engelmanii	95	A	1	0	2	3	0	0	0	1	1	0
Picea glauca	96	A	4	2	5	5	3	2	2	3	2	3
Picea glauca	97	B	0	0	0	2	0	0	0	0	0	0
Picea glauca	98	C	0	0	1	0	0	1	0	0	0	0
Picea mariana	99	A	0	0	0	0	0	0	0	0	0	0
Picea mariana	100	B	0	0	0	0	0	0	0	0	0	0
Picea mariana	101	C	0	0	0	0	0	0	0	0	0	0
Pinus contorta v. latifolia	102	A	0	0	0	0	0	0	0	0	0	1
Pinus contorta v. latifolia	103	B	0	0	0	0	0	0	0	0	0	0
Plagiothecium undulatum	104	D	0	1	0	0	0	0	0	0	0	0
Poa sp.	105	C	0	0	0	0	0	0	0	0	0	0
Polytrichum juniperinum	106	D	0	0	0	0	0	0	1	0	1	0
Populus X	107	A	0	0	0	0	0	0	0	0	0	0
Populus tremuloides	108	B	0	0	0	0	0	0	0	0	0	0
Populus tremuloides	109	A	0	0	0	0	0	0	0	0	0	0
Populus trichocarpa	110	A	0	0	0	0	0	0	0	0	0	0
Pseudotsuga menziesii	111	B	0	0	0	0	0	0	0	0	0	0
Pseudotsuga menziesii	112	A	0	0	0	0	0	0	0	0	2	1
Pteridium aquilinum	113	C	0	0	0	0	0	0	0	0	0	1
Pyrola asarifolia	114	C	0	1	0	0	0	0	0	0	0	0
Pyrola secunda	115	C	0	0	0	0	0	0	0	0	0	0
Ranunculus repens	116	C	0	0	0	0	0	0	0	0	0	0
Rhodobryum roseum	117	C	0	0	1	0	0	0	0	0	0	0
Rhytidiadelphus triquetus	118	D	1	0	1	1	1	1	0	1	0	0
Ribes hudsoniana	119	B	0	0	0	0	0	0	0	0	0	0
Ribes lacustre	120	B	1	1	2	1	1	0	0	1	0	1
Rosa sp.	121	B	0	0	0	1	0	0	0	0	1	0

Rubus ideus	122	B	0	0	0	1	0	1	0	0	0	1
Rubus parviflorus	123	B	1	1	0	0	0	0	0	1	1	2
Rubus pedatus	124	C	1	0	2	1	1	1	1	1	1	1
Rubus sp.	125	B	0	0	0	0	0	0	0	0	0	1
Salix sp.	126	B	0	0	0	0	0	0	0	0	0	0
Sambucus pubens	127	B	1	0	1	0	0	1	0	1	0	0
Sambucus sp.	128	B	0	0	0	0	0	0	0	0	0	0
Smilacina racemosa	129	C	1	1	1	0	1	1	0	1	1	0
Sonchus asper	130	C	0	0	0	0	0	0	0	0	0	0
Sorbus sitchensis	131	B	0	0	0	1	0	0	0	0	1	1
Sphagnum sp.	132	D	0	0	0	2	0	2	0	0	0	0
Spiraea densiflora	133	B	0	0	0	0	0	0	0	0	0	0
Spiraea lucida	134	B	0	0	0	0	0	0	0	0	1	1
Spiraea menziesii	135	B	0	0	0	0	0	1	0	0	0	0
Spiraea pyramidata	136	B	0	0	0	0	0	0	0	0	0	0
Sreptopus amplexifolius	137	C	1	1	1	1	0	1	0	1	0	0
Streptopus sp.	138	C	0	0	0	0	0	0	0	0	0	0
Streptopus roseus	139	C	1	1	0	1	1	1	0	1	1	1
Symphoricarpus albus	140	B	0	0	0	1	0	0	0	0	0	0
Thalictrum occidentale	141	C	0	0	0	0	0	0	0	0	1	0
Tiarella unifoliata	142	C	1	1	1	1	1	1	1	2	1	1
Trientalis latifolia	143	C	0	0	0	0	0	0	0	0	0	0
Tsuga heterophylla	144	A	0	0	0	0	0	0	0	0	0	0
Tsuga heterophylla	145	B	0	0	0	0	0	0	0	0	0	0
Urtica Lyalii	146	C	0	0	0	0	0	0	0	0	0	0
Vaccinium caespitosum	147	B	0	0	0	0	0	0	0	0	0	0
Vaccinium membraneaceum	148	B	0	0	1	1	0	1	0	0	0	1
Vaccinium occidentale	149	B	0	0	0	0	0	0	0	0	0	0
Vaccinium ovalifolium	150	B	0	0	0	0	0	0	0	0	0	0
Vaccinium oxicoccus	151	B	0	0	0	0	0	0	0	0	0	0
Vaccinium scoparium	152	B	0	0	0	0	0	0	1	0	0	0
Vaccinium vitis-idea	153	B	0	0	0	0	0	0	0	0	0	0
Veratrum viride	154	C	0	0	0	1	0	0	0	0	0	1
Viburnum parviflorum	155	B	1	0	1	0	0	0	0	1	0	1
Viola glabella	156	C	0	0	0	0	0	1	0	0	0	0

		Type 4					Type 5		
Taxa	#	1	2	3	4	5	1	2	3
Abies lasiocarpa	1	0	0	0	0	1	0	0	0
Abies lasiocarpa	2	1	0	1	1	1	0	0	0
Abies lasiocarpa	3	1	2	2	2	1	1	1	1
Acer glabrum	4	0	0	0	0	0	1	1	0
Aconitum columbianum	5	0	0	0	0	0	0	0	0
Actaea spicata	6	0	0	0	0	0	1	1	0
Agropyron sp.	7	0	0	0	1	0	1	0	0
Alnus synuata	8	0	0	0	0	0	0	0	0
Alnus tenuiflora	9	0	0	0	0	0	0	0	0
Alnus tenuifolia	10	0	0	0	0	0	0	0	0
Amelanchier alnifolia	11	1	1	1	1	1	1	1	1
Anaphalis margaretacea	12	0	0	0	0	0	0	0	1
Andromeda polyfolia	13	0	0	0	0	0	0	0	0
Aralia nudicaulis	14	1	1	1	1	1	1	2	3
Arnica cordifolia	15	0	0	0	1	0	0	0	1
Aruncus silvester	16	0	0	0	0	0	0	0	0
Asarum caudatum`	17	0	0	0	0	0	0	0	0
Aster douglasii	18	1	1	2	1	0	0	0	0
Aster sp.	19	0	0	0	0	0	0	0	0
Athyrium filix-femina	20	0	0	0	0	0	0	1	0
Betula glandulosa	21	0	0	0	0	0	0	0	0
Betula papyrifera	22	0	0	1	0	0	1	1	1
Betula papyrifera	23	0	0	0	0	0	0	0	0
Brachypodium sp.	24	0	0	0	0	0	1	0	0

Calamagrostis sp.	25	0	0	0	0	0	0	0	0
Carex lenticularis	26	0	0	0	0	0	0	0	0
Carex microglochin	27	0	0	0	0	0	0	0	0
Carex sp.	28	0	0	0	0	0	0	0	1
Carex tetanica	29	0	0	0	0	0	0	0	0
Carex trisperma	30	0	0	0	0	0	0	0	0
Chimaphylla umbellata	31	0	0	0	1	0	0	0	0
Chiogenes hispidula	32	1	1	1	1	2	0	0	0
Cinna latifolia	33	0	0	0	0	0	0	1	1
Circaea alpina	34	0	0	0	0	0	0	0	0
Cladonia rangiferina	35	1	0	0	1	0	0	0	0
Cladonia sp.	36	1	0	0	0	1	0	0	0
Cladonia sylvatica	37	1	1	1	0	1	0	0	0
Clintonia uniflora	38	1	1	1	1	1	1	1	0
Comandra umbellata	39	1	1	1	1	0	0	0	0
Comarum palustre	40	0	0	0	0	0	0	0	0
Cornus canadensis	41	1	2	1	1	1	1	2	1
Cornus stolonifera	42	0	1	0	1	0	1	1	1
Corylus cornuta	43	0	0	0	0	0	0	0	0
Dicranum scoparius	44	1	0	1	1	1	0	0	0
Disporum oreganum	45	0	0	1	0	0	1	0	0
Dryopteris phegopteris	46	0	0	0	0	0	0	0	0
Dryopteris austriaca	47	0	0	0	0	0	0	0	0
Dryopteris sp.	48	0	0	0	0	0	0	1	0
Dryopteris disjuncta	49	0	0	0	0	0	1	1	0
Dryopteris filix-mas	50	0	0	0	0	0	1	0	1
Elymus sp.	51	0	1	0	0	0	0	0	0
Entodon schreberii	52	1	2	1	1	4	0	0	0
Epilobium adenocaulon	53	0	0	0	0	0	0	0	0
Epilobium angustifolium	54	0	0	1	0	1	3	4	3
Equisetum fluviatilis	55	0	0	0	0	0	0	0	0
Equisetum pratense	56	0	0	0	0	0	1	0	1
Equisetum sylvaticum	57	0	0	0	0	0	1	0	1
Fragaria sp.	58	0	0	0	0	0	0	0	0
Fritillaria lanceolata	59	0	0	0	0	0	0	0	0
Galeopsis tetrachit	60	0	0	0	0	0	3	0	1
Galium aparine	61	0	0	0	0	0	1	0	0
Galium borealis	62	1	1	1	1	1	1	0	0
Galium triflorum	63	0	0	0	1	0	0	1	1
Geranium sp.	64	0	0	0	0	0	0	0	0
Geum oreganum	65	0	0	0	0	0	0	0	0
Habenaria sp.	66	1	0	0	0	0	0	0	0
Heracleum lanatum	67	0	0	0	0	0	0	0	0
Hylocomium splendens	68	3	2	1	1	2	0	0	0
Hypnum sp.	69	0	0	0	0	0	0	0	0
Impatiens biflora	70	0	0	0	0	0	0	0	0
Juniperus communis	71	0	1	0	0	0	0	0	0
Ledum groenlandicum	72	0	0	0	0	0	0	0	0
Larix laricina	73	0	0	0	0	0	0	0	0
Larix laricina	74	0	0	0	0	0	0	0	0
Lathyrus ochroleucus	75	1	1	0	1	1	0	1	0
Linnaea borealis	76	1	1	1	1	1	0	0	0
Listera cordate	77	0	0	0	0	0	0	0	0
Lonicera involucrata	78	0	0	1	0	0	1	1	0
Lycopodium annotinum	79	0	0	0	0	0	0	0	0
Lycopodium clavatum	80	0	0	0	1	1	0	0	0
Lycopodium complanatum	81	0	1	0	1	1	0	0	0
Lycopodium obscurum	82	1	0	1	0	0	0	1	0
Lyschitum americanum	83	0	0	0	0	0	0	0	0
Maianthemum uniflorum	84	1	1	1	0	0	0	0	0
Matteuccia struthiopteris	85	0	0	0	0	0	0	0	0
Melampyrum sp.	86	1	1	0	0	0	0	0	0
Mitella nuda	87	0	0	0	0	0	0	0	0
Mnium insigne	88	0	0	0	0	0	0	0	0
Mnium punctatum	89	0	0	0	0	0	0	0	0

Moneses uniflora	90	0	0	0	0	0	0	0	0
Oplopanax horridus	91	0	0	0	0	0	0	0	0
Orysopsis asperifolia	92	1	1	1	1	1	0	0	0
Peltigera sp.	93	1	1	1	1	2	0	0	0
Petasites speciosa	94	1	1	1	1	1	0	0	0
Picea engelmanii	95	0	0	0	0	0	0	0	0
Picea glauca	96	0	0	0	0	0	0	0	0
Picea glauca	97	1	0	0	0	0	0	0	0
Picea glauca	98	1	0	1	0	0	0	0	1
Picea mariana	99	2	2	2	2	3	0	0	0
Picea mariana	100	0	2	2	1	1	0	0	0
Picea mariana	101	1	0	1	1	1	0	0	0
Pinus contorta v. latifolia	102	2	2	2	2	1	0	0	0
Pinus contorta v. latifolia	103	0	0	0	1	0	0	0	0
Plagiothecium undulatum	104	0	0	0	0	0	0	0	0
Poa sp.	105	0	0	0	0	0	1	0	0
Polytrichum juniperinum	106	0	0	0	0	0	0	0	0
Populus X	107	0	0	1	0	0	0	0	0
Populus tremuloides	108	0	0	1	0	0	0	0	0
Populus tremuloides	109	0	0	1	0	0	1	5	1
Populus trichocarpa	110	0	0	0	0	0	0	0	0
Pseudotsuga menziesii	111	1	0	1	1	1	0	0	0
Pseudotsuga menziesii	112	3	1	2	1	1	0	0	0
Pteridium aquilinum	113	0	0	0	0	0	0	0	0
Pyrola asarifolia	114	0	0	0	0	0	0	0	0
Pyrola secunda	115	1	0	0	0	0	0	0	0
Ranunculus repens	116	0	0	0	0	0	0	0	0
Rhodobryum roseum	117	0	0	0	0	0	0	0	0
Rhytidiadelphus triquetus	118	0	1	0	0	0	0	0	0
Ribes hudsoniana	119	0	0	0	0	0	0	0	0
Ribes lacustre	120	0	0	0	0	0	1	1	1
Rosa sp.	121	1	1	1	1	1	1	1	1
Rubus ideus	122	0	0	0	0	0	1	1	0
Rubus parviflorus	123	0	1	0	1	1	3	3	3
Rubus pedatus	124	1	1	1	1	0	0	0	0
Rubus sp.	125	1	0	1	0	0	1	1	0
Salix sp.	126	0	0	0	0	0	2	0	1
Sambucus pubens	127	0	0	0	0	0	0	0	0
Sambucus sp.	128	0	0	0	0	0	1	0	0
Smilacina racemosa	129	1	1	1	1	0	1	1	1
Sonchus asper	130	0	0	0	0	0	0	0	0
Sorbus sitchensis	131	0	0	0	0	0	1	0	1
Sphagnum sp.	132	0	0	0	0	0	0	0	0
Spiraea densiflora	133	0	0	0	0	0	0	0	0
Spiraea lucida	134	1	1	1	1	1	0	0	0
Spiraea menziesii	135	0	0	0	0	0	0	0	0
Spiraea pyramidata	136	1	1	1	1	1	0	0	0
Sreptopus amplexifolius	137	0	0	0	0	0	1	0	0
Streptopus sp.	138	1	1	1	0	1	1	0	0
Streptopus roseus	139	0	0	0	0	0	0	0	0
Symphoricarpus albus	140	0	0	0	0	0	1	0	1
Thalictrum occidentale	141	0	0	0	0	0	1	0	1
Tiarella unifoliata	142	0	0	0	0	0	1	1	1
Trientalis latifolia	143	0	0	0	0	0	0	0	0
Tsuga heterophylla	144	0	0	0	0	0	0	0	0
Tsuga heterophylla	145	1	0	1	1	1	0	0	0
Urtica Lyalii	146	0	0	0	0	0	0	0	0
Vaccinium caespitosum	147	1	3	2	2	1	0	0	0
Vaccinium membraneaceum	148	2	1	1	1	1	1	1	0
Vaccinium occidentale	149	2	2	1	2	1	0	0	0
Vaccinium ovalifolium	150	0	1	1	0	0	0	0	0
Vaccinium oxicoccus	151	0	0	0	0	0	0	0	0
Vaccinium scoparium	152	0	0	0	0	0	0	0	0
Vaccinium vitis-idea	153	0	0	0	0	0	0	0	0
Veratrum viride	154	0	0	0	1	0	0	0	1

| Viburnum parviflorum | 155 | 1 | 1 | 1 | 1 | 1 | 1 | 1 | 1 |
| Viola glabella | 156 | 1 | 1 | 0 | 0 | 0 | 1 | 0 | 0 |

Notes:

www.ingramcontent.com/pod-product-compliance
Lightning Source LLC
Chambersburg PA
CBHW061219180526
45170CB00003B/1064